Analog Circuits and Signal Processing

Series Editors
Mohammed Ismail, Khalifa University, Dublin, OH, USA
Mohamad Sawan, Montreal, QC, Canada

The *Analog Circuits and Signal Processing* book series, formerly known as the *Kluwer International Series in Engineering and Computer Science*, is a high level academic and professional series publishing research on the design and applications of analog integrated circuits and signal processing circuits and systems. Typically per year we publish between 5–15 research monographs, professional books, handbooks, and edited volumes with worldwide distribution to engineers, researchers, educators, and libraries.

The book series promotes and expedites the dissemination of new research results and tutorial views in the analog field. There is an exciting and large volume of research activity in the field worldwide. Researchers are striving to bridge the gap between classical analog work and recent advances in very large scale integration (VLSI) technologies with improved analog capabilities. Analog VLSI has been recognized as a major technology for future information processing. Analog work is showing signs of dramatic changes with emphasis on interdisciplinary research efforts combining device/circuit/technology issues. Consequently, new design concepts, strategies and design tools are being unveiled.

Topics of interest include:
Analog Interface Circuits and Systems;
Data converters;
Active-RC, switched-capacitor and continuous-time integrated filters;
Mixed analog/digital VLSI;
Simulation and modeling, mixed-mode simulation;
Analog nonlinear and computational circuits and signal processing;
Analog Artificial Neural Networks/Artificial Intelligence;
Current-mode Signal Processing;
Computer-Aided Design (CAD) tools;
Analog Design in emerging technologies (Scalable CMOS, BiCMOS, GaAs, heterojunction and floating gate technologies, etc.);
Analog Design for Test;
Integrated sensors and actuators;
Analog Design Automation/Knowledge-based Systems;
Analog VLSI cell libraries;
Analog product development;
RF Front ends, Wireless communications and Microwave Circuits;
Analog behavioral modeling, Analog HDL.

More information about this series at https://link.springer.com/bookseries/7381

Nourhan Elsayed • Hani Saleh • Baker Mohammad
Mohammed Ismail • Mihai Sanduleanu

High Efficiency Power Amplifier Design for 28 GHz 5G Transmitters

Nourhan Elsayed
Khalifa University
Abu Dhabi, UAE

Baker Mohammad
Khalifa University
Abu Dhabi, UAE

Mihai Sanduleanu
Khalifa University
Abu Dhabi, UAE

Hani Saleh
Khalifa University
Abu Dhabi, UAE

Mohammed Ismail
Wayne State University
Detroit, MI, USA

ISSN 1872-082X ISSN 2197-1854 (electronic)
Analog Circuits and Signal Processing
ISBN 978-3-030-92748-6 ISBN 978-3-030-92746-2 (eBook)
https://doi.org/10.1007/978-3-030-92746-2

This Springer imprint is published by the registered company Springer Nature Switzerland AG
The registered company address is: Gewerbestrasse 11, 6330 Cham, Switzerland

For my parents, Somaya and Mohamed.

Nourhan Elsayed

Preface

The wireless communication market has witnessed a substantial level of development and growth in recent years. The ever-increasing number of subscribers, demand for higher data rate, and better connectivity has driven the markets to provide more efficient communication schemes with a low cost-to-market. The available mm-wave spectrum has driven work towards introducing 5G technology. 5G technology is expected to operate in the 28 GHz band with high peak-to-average power ratio (PAPR). This is a consequence of orthogonal frequency division multiplexing (OFDM) modulation combined with QAM modulated OFDM carrier signals that enable high data rates and high spectral efficiency.

The power amplifier (PA) is a key building block in a transmitter and is the most power-hungry component in modern systems. To offer lower cost of integration, it is desirable to have the entire transceiver system, including the PA, on a single CMOS chip. Some of the major challenges that CMOS power amplifiers face at mm-wave frequencies are limited output power per transistor and low power added efficiency. The recent availability of scaled CMOS technology with cut-off frequency larger than 300 GHz has opened the possibility of implementing mm-wave systems at a high level of integration. Despite that, stringent requirements on the PA to achieve the desired efficiency, linearity, and output power while operating at lower supply voltages still impose a challenge. This has driven a lot of research towards different circuit techniques, and design trade-offs for power amplifiers in scaled CMOS technologies. However, there is still a clear gap in achieving the desired performance in PAs dedicated towards 5G applications.

The first part of this book presents the design of highly efficient PA architectures in 22 nm FDSOI technology at 28 GHz. This effort was done in order to achieve high efficiency while maintaining the stringent linearity and output power requirements imposed by high PAPR signals. Four different PA architectures were designed: one based on linear PAs, two switching PAs, and one hybrid. All these PAs were fabricated in GlobalFoundries and measurement results are presented and compared with simulation. The first PA presented is based on a Doherty architecture with a Class-A PA as the main amplifier, and Class-C as the auxiliary amplifier, both utilizing the stacking technique to overcome the low breakdown voltage of the devices.

A switched-mode Class-E PA was also designed utilizing the cascode topology. A novel technique of switching the cascode device along with the input device, referred to as *pulse injection*, has been employed to maximize efficiency. The PA achieves peak measured power added efficiency (PAE) of 35%, drain efficiency (DE) of 45%, and 8.5 dB power gain. While this PA achieves high efficiency, linearity is still a concern. As a means of linearization, a Class-E based Doherty PA was designed in order to utilize the high efficiency achieved from the auxiliary Class-E PA while maintaining linearity due to the Doherty configuration. The Doherty-based Class-E PA achieves a measured peak PAE of 32% and 31% at 6-dB back-off and hence maintains the efficiency at high and backed-off power levels, with a gain of 17 dB.

Next, a current mode Class-D (CMCD) PA utilizing the cascode topology was designed. The CMCD PA also utilizes pulse injection to switch the cascode device via a tunable transmission line as a delay element. Measurement results show a measured peak PAE of 46% and drain efficiency (DE) of 71% with a saturated output power of 19 dBm. This PA reports the best performance compared to other CMCD current mode Class-D PAs in literature at mm-waves.

Finally, the measured Class-E Doherty-based PA is integrated into a 4 phased-array transmitter design utilizing a novel active power divider. Measurement results of the stand-alone power divider were used to validate its functionality. Simulation results are presented for the full transmitter chain. Currently, a PCB is being developed to measure and validate the 4-phased-array transmitter design.

All in all, with the speeding race to offer 5G communication scheme to the market, this book aims to fill the gap in research between achieving high efficiency and a high level of integration through utilizing scaled CMOS technology (22 nm FDSOI).

Abu Dhabi, United Arab Emirates   Nourhan Elsayed

Abu Dhabi, United Arab Emirates   Hani Saleh

Abu Dhabi, United Arab Emirates   Baker Mohammad

Detroit, United States of America   Mohammed Ismail

Abu Dhabi, United Arab Emirates   Mihai Sanduleanu

Acknowledgments

This research was sponsored by GlobalFoundries (GF) and Semiconductor Research Corporation (SRC) through a grant to the System-on-Chip Center at Khalifa University. The team from GF and SRC was always hands-on with our work, providing industry insight, feedback, and technology support. None of this would have been possible without your efforts. Special thanks to Dan Cracan and Ademola Mustapha for their support and contribution throughout the research.

Contents

Acronyms

BO	Back Off
BPF	Band Pass Filter
CMCD	Current Mode Class D
CMOS	Complementary Metal Oxide Semiconductor
CPW	Co Plannar Waveguide
DAC	Digital to Analog Converter
DC	Direct Current
DE	Drain Efficiency
DPA	Doherty Power Amplifier
DUT	Device Under Test
EE	Energy Efficiency
EIRP	Equivalent Isotropically Radiated Power
ET	Envelope Tracking
FBB	Forward Body Bias
FDSOI	Fully Depleted Silicon On Insulator
FoM	Figure of Merit
GSG	Ground-Signal-Ground
IOT	Internet of Things
LF	Low Frequency
LO	Local Oscillator
LPF	Low Pass Filter
MIMO	Multiple Input Multiple Output
NF	Noise Figure
OFDM	Orthogonal Frequency Division Multiplexing
PA	Power Amplifier
PAE	Power Added Efficiency
PAPR	Peak to Average Power Ratio
PWM	Pulse Width Modulation
QAM	Quadrature Amplitude Modulation
QOS	Quality of Service
RBB	Reverse Body Bias

RF	Radio Frequency
RFC	Radio Frequency Choke
RFIC	Radio Frequency Integrated Circuit
SE	Spectral Efficiency
SMPA	Swicthed Mode Power Amplifier
SNR	Signal to Noise Ratio
TL	Transmission Line
VGA	Variable Gain Amplifier
VMCD	Voltage Mode Class D
VNA	Vector Network Analyzer
ZCS	Zero Current Switching
ZVS	Zero Voltage Switching

Chapter 1
Introduction

In recent years, there has been a drastic increase in the use of smart devices, the number of users, and a great demand in downloading multimedia content. This increase can no longer be supported by the current 4G technology (Fig. 1.1). Mobile users are expecting more reliability and a faster service leading to the introduction of 5G as the future of communication [2]. 5G technology will not only impact mobile services, but also the Internet of Things (IoT), autonomous vehicles, and smart cities [3]. 5G technology is expected to achieve [4, 5]:

- 100x more capacity
- Higher data rate (up to 1 Gb/s)
- Lower end-to-end latency (<1 ms)
- Better quality of service (QOS)
- Massive device connectivity

The limitation of the current RF band used by the service providers is that it is over-crowded with the high data consumption of the current users. Therefore, 5G technology will utilize a new spectrum that has never been used by mobile services before. While 4G currently operates between 1.8 and 2.6 GHz, 5G spectrum is in the sub-6-GHz range and the millimeter-wave (mm-Wave) frequency range (>24.25 GHz) [4]. The lower frequency bands will be used for less-densely populated areas because data can travel longer distances. The higher mm-Wave frequency range shows prospect of supporting hundreds of times more data rate and capacity over the current cellular spectrum opening up a new horizon for spectrum constrained future wireless communications [2, 3].

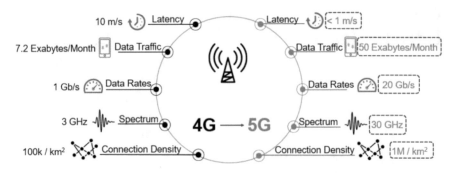

Fig. 1.1 5G capabilities compared to previous communication standards [1]

1.1 Motivation for High Efficiency Power Amplifiers for 5G Technology

5G technology is expected to operate in the 28 GHz band with high Peak to Average Power Ratio (PAPR). This is a consequence of OFDM modulation combined with QAM modulated OFDM carrier signals such as 64 and 128 QAM that offer high data rate and spectral efficiency (SE). The trade-off this presents is that we need to have high energy efficiency (EE), with a high quality of service (QoS), while maintaining a high spectral efficiency (SE). These requirements are sometimes conflicting for a transmitter to maintain considering the higher Bandwidth requirement [5] (Fig. 1.2).

The Power Amplifier (PA) is a key building block in a transmitter and is the most power-hungry component in modern systems. To offer lower cost of integration, it is desirable to have the entire transceiver system, including the PA, on a single CMOS chip. Some of the major challenges that CMOS power amplifiers face at mm-Wave frequencies are limited output power per transistor and low power added efficiency. The recent availability of scaled CMOS technology with cut-off frequency larger than 300 GHz has opened the possibility of implementing mm-Wave systems at a high level of integration. Despite that, stringent requirements on the PA to achieve the desired efficiency, linearity, and output power while operating at lower supply voltages still impose a challenge. This has driven a lot of research toward different circuit techniques and design trade-offs for power amplifiers in scaled CMOS technologies. However, there is still a clear gap in achieving the desired performance in PAs dedicated toward 5G applications.

In a traditional transmitter shown in Fig. 1.3, the power amplifier is the most power-hungry component (50–80%) that determines the efficiency, cost, and overall performance of the system [6]. The design of the power amplifier becomes critical. Due to large PAPR, it is difficult to design a power amplifier with high efficiency, high power gain, and high linearity.

Modern scaled CMOS technologies are exhibiting high cut-off frequencies $(f_T > 300)$ GHz that allow for mm-wave operation. However, challenges such

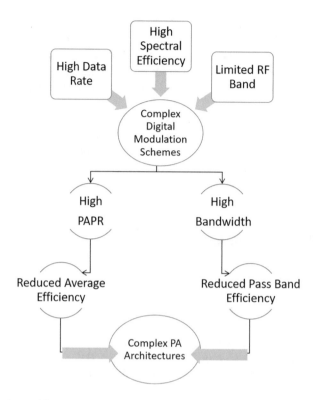

Fig. 1.2 The demand for advanced power amplifier architectures

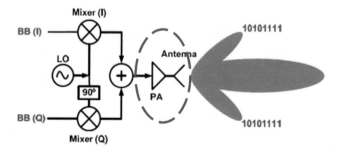

Fig. 1.3 Traditional analog I/Q transmitter architecture

as drain–source breakdown voltage, limited output power capability per transistor, and poor passive element quality factor still present an obstacle. High power devices such as SiGe, GaAs, and GAN are still favorable due to their high output power and high efficiency capabilities at frequencies up to 90 GHz. While these technologies offer high performance, they are costly to implement on a large scale and do not offer the flexibility of high levels of integration.

This is providing motivation for researchers to develop circuit techniques at lower cost technologies such as CMOS that also provides high levels of integration to drive the future commercialization of high data rate communication systems. New Silicon-On-Insulator (SOI) and Fully Depleted Silicon On Insulator (FDSOI) technologies are opening the door to CMOS becoming a more promising alternative to III–V technologies.

The main goal of this book is to set stepping stones toward circuit techniques specifically dedicated toward the design of high efficiency power amplifiers for 5G transmitter applications for handset and cellular mobile device applications. The proposed work utilizes the advantages of deeply scaled CMOS technologies (22nm FDSOI) to push the boundaries of the existing state-of-the-art performance.

1.2   Book Organization

This book starts with Chap. 2, which provides basic definitions of power amplifier fundamentals starting with performance measures for efficiency and linearity. The characterization of different power amplifier classes and challenges is also discussed followed by different techniques in the literature to overcome them.

Chapter 3 presents the design of a 28 GHz Doherty power amplifier. The chapter discusses the design methodology including transistor sizing and circuit design. This is followed by simulation and measurement results from the fabricated chip. Finally, the results are compared with the existing state-of-the-art CMOS Doherty power amplifiers.

Chapter 4 presents the design of a novel switched cascode Class-E power amplifier at 28 GHz. The novel technique of switched cascode is discussed, which utilizes a tunable transmission line. On-chip measurement results including small and large signal measurements are presented. Finally, the comparison with the state-of-the-art Class-E amplifiers is discussed where the designed Class-E PA out outperforms the state-of-the-art designs that can be observed from the figure of merit.

Chapter 5 presents the integration of the switched cascode Class-E PA into a Doherty architecture. The architecture makes use of the high efficiency Class-E PA while maintaining it at back-off due to the Doherty configuration. The improved Doherty PA is compared with the classical Doherty, and the improvement in performance is observed.

Chapter 6 presents the design of cascode inverse Class-D power amplifier that utilizes pulse injection via a tunable transmission line. The design methodology is presented. Simulation and on-chip measurement results are provided along with a comparison with the existing inverse Class-D amplifiers.

Chapter 7 presents the design of a 4-phased-array transmitter topology. Transmitter includes the blocks needed from the baseband signal to the RF. It utilizes a novel power divider and tunable transmission lines as delay elements. Each block design is discussed, along with simulation and some on-chip measurement results.

References

1. M. Mageean, Is 5G the future of mobile technology [Online]. http://www.verifyrecruitment.com/blog/index.php/5g-the-future-of-mobile-technology/
2. M. Agiwal, A. Roy, N. Saxena, Next generation 5G wireless networks: a comprehensive survey. IEEE Commun. Surv. Tuts. **18**(3), 1617–1655 (2016)
3. S. Buzzi, C.L. I, T.E. Klein, H.V. Poor, C. Yang, A. Zappone, A survey of energy-efficient techniques for 5G networks and challenges ahead. IEEE J. Sel. Areas Commun. **34**(4), 697–709 (2016)
4. P.M. Asbeck, Will Doherty continue to rule for 5G? in *2016 IEEE MTT-S International Microwave Symposium (IMS)* (2016), pp. 1–4
5. A. Gupta, R.K. Jha, A survey of 5G network: architecture and emerging technologies. IEEE Access **3**, 1206–1232 (2015)
6. J. Joung, C.K. Ho, K. Adachi, S. Sun, A survey on power-amplifier-centric techniques for spectrum- and energy-efficient wireless communications. IEEE Commun. Surv. Tuts. **17**(1), 315–333 (2015)

Chapter 2
Power Amplifier Fundamentals

The transmitter in an RF system is responsible for transmitting RF signals modulated with digital information to the receiver via wireless means. The power amplifier in a transmitter is usually the most power consuming part since it converts DC power into RF power. Therefore, the competence of a transmitter is directly affected by the power amplifier. When operating at a high frequency, challenges such as linearity, especially with a high Peak to Average Power (PAPR) ratio, become prominent. Factors such as the breakdown voltage of the devices, the operating frequency, and the losses in the transmission medium have a significant effect on the performance of the amplifier. Various techniques have been developed in order to improve the efficiency of RF power amplifiers. In this chapter, different performance metrics of the power amplifier will be discussed along with different classes of PAs including saturated and unsaturated. Different architectures in the literature will be compared alongside technology consideration and its effect on the overall system.

2.1 Performance Metrics

2.1.1 Power Amplifiers Efficiency

Efficiency is the one of the most important measures of a power amplifier and greatly affects the whole efficiency of the transmitter. The more efficient the power amplifier is, the less power consumption is needed by the transmitter. It is estimated to consume 70% of the power used in a transmitter chain. The benefits of a highly efficient transmitter are a longer battery life in handheld devices and lower costs for cooling. There are a number of factors that affect the efficiency of an amplifier:

- DC power

© The Author(s), under exclusive license to Springer Nature Switzerland AG 2022

N. Elsayed et al., *High Efficiency Power Amplifier Design for 28 GHz 5G Transmitters*, Analog Circuits and Signal Processing,
https://doi.org/10.1007/978-3-030-92746-2_2

- Operating frequency
- Design architecture
- Load

There are two main efficiency measures that are used commonly as figures of merit for RF PAs, the power added efficiency (PAE), and the drain efficiency (DE) [1]. The drain efficiency is the ratio of the average output power and the DC power consumed by the PA. Drain efficiency can be calculated by

$$DE = \frac{P_{out}}{P_{DC}} \tag{2.1}$$

where

$$P_{DC} = V_{DD} * I_{DC} \tag{2.2}$$

In other words, the drain efficiency calculates how much DC power is consumed to amplify the input power to the output power. Another widely used method of measuring efficiency is the power added efficiency (PAE) that includes the input power into the calculation:

$$PAE = \frac{P_{out} - P_{in}}{P_{DC}} \tag{2.3}$$

This provides another measure to quantify the power conversion ability of the power amplifier that includes the input RF power (P_{in}).

2.1.2 Output Power Capability

The power amplifier needs to be able to deliver the output power required for the communication system. The output power (P_{out}) is generally measured in dBm. The maximum output power that the power amplifier can deliver is also the saturated power of the amplifier (P_{sat}). In an RF transmitter, a purely resistive load is assumed at the frequency of interest, and the antenna can be represented as a resistor having the value of 50 Ω as shown in Fig. 2.1. The load impedance can be transformed into the desired value with an imaginary part by using a matching network. In this research our load is transformed into 50 Ω in order to achieve maximum power delivery [2]. In a PA, the instantaneous output power at a particular time instant can be defined as

$$P_{out,inst}(t) = v_{out}(t) * i_{out}(t) \tag{2.4}$$

The average power delivered to the load is

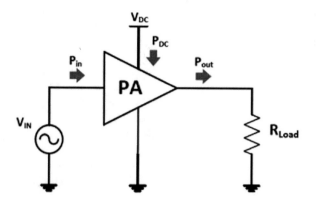

Fig. 2.1 PA output power calculations assuming a resistive load

$$P_{out,avg} = \lim_{T \to \infty} \int_{-T/2}^{T/2} P_{out,inst}(t)dt \qquad (2.5)$$

Therefore, the average output power at the fundamental frequency (f_0) when the amplifier is operating in linear mode or small signal conditions is

$$\frac{V_{max}}{2R_L} = \frac{V_{rms}^2}{R_L} \qquad (2.6)$$

From the equations, we can notice that the output power is directly dependent on the supply voltage [3]. Therefore, it would make sense to increase it as much as possible in order to deliver more output power. However, with CMOS technology, supply voltage cannot be increased infinitely because the devices have a breakdown voltage and it results in decreased device reliability.

2.1.3 Linearity

With new modulation schemes aiming to enhance higher spectral efficiency through utilizing a high PAPR, linearity requirements for the PA become more essential. Non-linearities in a PA lead to spectral regrowth and hence interference from "near-by" channels and result in amplitude compression. These effects can be quantified through some indicators such as 1-dB compression point, third order intermodulation distortion, and third order intercept point (IP3).

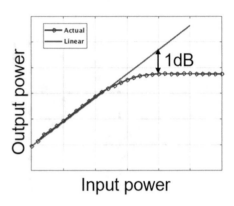

2.1.3.1   1-dB Compression Point

1-dB compression point (AM–AM conversion) studies the non-linearity in the amplitude relationship between the input and output signals with P_{1-dB} as a metric. Figure 2.2 shows a plot of the theoretical linear relationship between the output and input signal amplitudes and the actual ones. The point at which the difference between the two is exactly 1 dB is the P_{1-dB}. This point shows the maximum input power level that the PA can handle without losing linearity and resulting in other spectral components [2]. With high PAPR signal it is desired to have a higher P_{1-dB} because it means that the PA can withhold its linearity within the range of input amplitude. However, this also means that the PA is operating mostly in the low input power region, which, while maintaining a high gain, sacrifices efficiency.

2.1.3.2   Third Order Intercept Point (IP3)

Third order intermodulation point provides a good metric when dealing with a system with multiple RF blocks since it is independent of the input power levels. To calculate the input inferred third order intercept point (IIP3), two interferers close to the operating frequency are assumed, which will result in a non-linear output known as the intermodulation product. The intermodulation product of interest is the third order since it falls in the desired frequency band. The amplitude of the third order intermodulation product is cubed of the fundamental amplitude. Hence, it can increase to be as large as the fundamental after a certain input power level. The third order interception point is the extrapolated intersection of the curves of the fundamental and the third order power. This way, the input/output referred IP3 can be estimated.

2.1.4   Figure of Merit (FoM)

For a fair more inclusive comparison between PAs and their performance, a figure of merit (FoM ITRS) was introduced by [4] that includes the main performance metrics and the limitations of a PA including the saturated power (P_{sat}), the gain, the PAE, and the ratio of the operating frequency f_0 compared to the f_{max} of the device, defined as

$$\text{FoM} = P_{sat} \text{ (dBm)} + \text{Gain (dB)} + 10 \log_{10} \text{(PAE)}$$
$$+ 20 \log_{10} \frac{f_0}{f_{max}} \quad\quad (2.7)$$

This FoM is a modification from a previous FoM mentioned in [5], which only accounts for the operating frequency irrespective of the f_{max} of the device.

2.2   Power Amplifier Classes

Power amplifiers can be mainly classified into two main modes of operation: linear and switching amplifiers. For linear amplifiers, the output power is a linear function of the input power. Such amplifiers are Class-A,B, and A/B. For switching amplifiers, instead of the transistor operating as a current source, it operates as an ON/OFF switch depending on the level of the input signal. These amplifiers can provide 100% theoretical efficiency but compromise the linearity of the PA.

2.2.1   Linear Power Amplifiers (A, A/B, B, and C)

Classes A, AB, B, and C amplifiers utilize a similar circuit configuration and are separated by the biasing conditions. The active device acts as a transconductance (g_m) that converts the input voltage into current and passes it to the load. These amplifiers are called linear amplifiers because they linearly amplify the components at the fundamental frequency preserving both phase and amplitude [2]. The device is only active for a portion of the input signal referred to as the conduction angle. Class-A amplifier has the largest conduction angle (360 °) and is the simplest one; however, it suffers from having the least linearity. The drain efficiency of the PA can be calculated by

$$DE(\eta) = \frac{P_{out}}{P_{DC}} = \frac{i_d^2 R_L}{2 V_{DD} i_d} \quad\quad (2.8)$$

From this equation we can deduce that the maximum theoretical drain efficiency is only 50%, which decreases when implemented with real components.

In order to minimize the loss in instantaneous power, the conduction angle can be lowered, which gives rise to Class-AB (180° < conduction angle <360°), Class-B (conduction angle= 180°), and Class-C (conduction angle< 180°). This way, the efficiency increases with decreasing the conduction angle. Although we can increase the efficiency by moving to Class-C and reducing the conduction angle, we are compromising the output power and gain.

2.2.2 Switched Mode Power Amplifiers (SMPAs)

As a way to improve a PA's efficiency, the overlapping time of output voltage and current waveforms needs to be reduced. This is where switching PAs come into play. Instead of the devices acting as a transconductance, they are employed as switches. Therefore, the device is either ON with a large input signal and OFF otherwise, therefore theoretically removing power losses and achieving 100% efficiency. Switching PA classes include Class-D and Class-E that are widely used in handheld devices. Table 2.1 provides a comparison between different switching mode PA classes and their output power capabilities.

2.2.3 Summary

In summary, there is always a trade-off between linearity and efficiency when determining the mode of operation of a PA. Table 2.2 summarizes power amplifier classifications along with the efficiency and linearity.

Table 2.1 Comparison of different switched mode PA classes

Class	DE	Peak drain voltage	Peak drain current	Max. output power	Power[a] capability
D	100	V_{DD}	$\frac{V_{DD}}{R}$	$\frac{(2V_{DD})^2}{\frac{\pi}{2R}}$	0.32
D^{-1}	100	πV_{DD}	$\frac{\pi V_{DD}}{R}$	$\frac{(\pi V_{DD})^2}{2R}$	–
E	100	$3.6V_{DD}$	$\frac{1.7V_{DD}^2}{R}$	$\frac{0.577V_{DD}^2}{R}$	0.098
F	100	$\frac{8V_{DD}}{\pi}$	$\frac{8V_{DD}}{\pi R}$	$\frac{4V_{DD}^2}{2R}$	0.16

[a] Ratio of actual output power to the product of the maximum device voltage and current

Table 2.2 Summary and comparison of power amplifier classes

Class	Mode	Conduction angle	Efficiency (%)	Linearity	Gain
A	Linear	100	50	High	High
B		50	78.5	Intermediate	Intermediate
C		<50	100	Low	Low
D	Switching	50	100	Low	Low
E		50	100	Low	Low
F		50	100	Low	Low

2.3 Efficiency Enhancement Techniques

Advanced modulation schemes utilizing high PAPR call for power amplifiers with high linearity performance. However, PAs suffer from low efficiency at backed-off power levels where the PA is operating 90% of the time. More efficient PAs such as switching PAs would deliver higher efficiency but at the expense of linearity. Improving the efficiency of PAs while not sacrificing linearity has been the topic of wide efforts. The next sections discuss some of the most common techniques in the literature.

2.3.1 Dynamic Biasing

As discussed in the previous sections, the efficiency of highly linear PAs drops at low input signal levels. Therefore, the PA cannot maintain efficiency at both low and high signal power levels. One way to overcome this is to utilize dynamic biasing, where the DC bias voltage of the device varies with the input power level. At low input amplitudes, the bias voltage decreases, the DC current drops, and the mean efficiency increases and vice versa. The disadvantages of this method include high distortion and drops in power gain.

2.3.2 Envelope Tracking

The first CMOS amplifier used for wireless communication is the Envelope Tracking (ET) power amplifier used to improve the power added efficiency of wireless transmitters [6]. The ET PA operates in such a way that the voltage applied to the PA is continuously adjusted to make sure that it is always operating at its peak maximum power and peak efficiency. Compared to other classical PAs that operate below their peak power level and maximum efficiency (compression). One of the dominant limitations in an ET PA is the limited signal bandwidth (20–40 MHz) that does not match the movement toward higher frequencies for wireless access methods [7].

2.3.3 Doherty Configuration

One of the most popular configurations for enhancing efficiency for PAs dedicated toward 5G is the Doherty PA shown in Fig. 2.3. This amplifier is a linear power amplifier that combines two linear PAs through a combiner. It has high efficiency and linearity over a wide frequency band and power levels [8]. A Doherty amplifier can increase efficiency from 11% to 14% when compared to the standard parallel Class-AB amplifiers.

The Doherty PA is appealing for 5G technology due to the following aspects [10]:

- Improved power added efficiency, compared to Class-A/AB amplifiers, at lower power levels, in back-off.
- High efficiency and linearity over wide frequency band and power levels.
- Only needs a quarter-wave transmission line to combine two output powers. Those can fit, on-chip, at high frequencies.

The Doherty power amplifier (DPA) combines two linear PAs. The first amplifier is the main amplifier. The other amplifier referred to as the auxiliary amplifier turns on when the input power increases and as the main amplifier runs into compression. This process enables accommodating more output power at low and high signal power levels. One of the most important aspects of the operation of the Doherty PA is to get the auxiliary PA to operate when the main runs into compression. If it operates at all power levels, the overall efficiency will be compromised.

The signal enters the amplifier through a splitter that creates two signals 90° out of phase with respect to each other. Such splitters have a typical 3-dB power loss as the price for creating a phase shift. The signal then goes through both amplifiers (main and auxiliary). At low input power levels, only the main PA is active, while at higher power levels, the auxiliary PA comes into play. Hence, it exhibits high

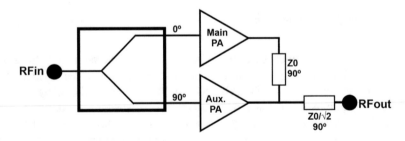

Fig. 2.3 Classical Doherty configuration [9]

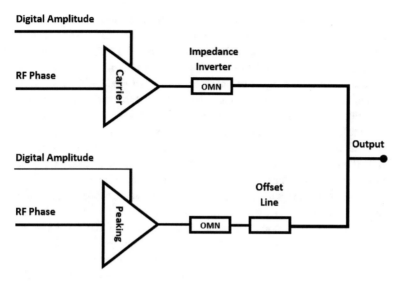

Fig. 2.4   Digital polar Doherty PA

efficiency over the whole input power range. After the signals pass through the amplifiers, they are combined again through reversing the operation of the splitter circuit at the input. This is achieved by adding another 90° phase shift through a quarter wavelength transmission line. As a result, the signals from both amplifiers remain in phase. Since both amplifiers are matched to $Z_0 = 50\,\Omega$, another quarter-wave transformer is added at the output as an impedance transformer.

In some cases, both amplifiers can operate in the same class. However, adaptive biasing techniques need to be applied in order to turn on/off the auxiliary amplifier when required. This complicates the design with extra circuitry to sense the input power levels and the need to turn on the auxiliary amplifier at the right power level of the input. There has been extensive research in utilizing the advantages of the Doherty architecture in improving efficiency at peak and back-off power levels. A review of the existing DPAs has been done. The first DPA presented is a digital polar Doherty power amplifier shown in Fig. 2.4 implemented in 180 nm CMOS [11]. The PA is digitized since it has a 10 bit amplitude control for each amplifier (main and auxiliary) facilitating adaptive biasing. It gives off 33.1 dBm peak power at 55.5% efficiency and a 52.5% efficiency at 6 dB back-off. The problem with this amplifier is that it operates at a frequency of 0.9 GHz, which is much less than the frequency range of 5G. It also requires a combiner that is off-chip and hence cannot be applied to a fully integrated chip.

The next Doherty PA presented is implemented in 45 nm CMOS and operates at a frequency of 45 GHz [12, 13]. It utilizes two-stack amplifiers and achieves 18 dBm peak power at 23% efficiency and 17% efficiency at 6 dB back-off. The amplifier is fully integrated at an area of 0.636 mm². The amplifiers are stacked in order to overcome the low breakdown voltage issue and to achieve high output voltage. Impedance matching is utilized at intermediate nodes for maximum gain; however, the gain reported is 6 dB, which is relatively low for 5G applications.

Another example is a 60 GHz DPA implemented in 0.13 μm utilizing microstrip transmission lines for impedance matching presented in [14]. The amplifier achieves the best reported figure of merit for fully integrated CMOS mm-Wave PAs, which links the output power with the gain. While it achieves a peak power gain of 13.5 dB, but efficiency performance degrades to 3% PAE.

While the Doherty power amplifier provides a technique to enhance efficiency at backed-off power levels and hence improving the overall performance, it is still limited to the maximum efficiency of the auxiliary amplifier (Class-C). Switching amplifiers, on the other hand, can achieve a theoretical efficiency of 100%. Some classes of switching PAs such as Class-E have been of particular interest due to its relatively simple architecture and matching networks such as the work presented in [4, 5, 15]. However, this comes at the cost of linearity and switching losses. In [4, 5], the stacking technique is used to overcome the breakdown voltage of the devices and deliver higher power. While the peak PAE achieved is higher than that of the linear counterparts (34% in [5]), the degradation at backed-off power levels is still steep.

2.4   Technology Limitations

2.4.1   Low Breakdown Voltages

The breakdown of the gate oxide occurs when the electric field in the oxide exceeds a certain value that is technology specific. This results in destroying the transistor due to the short circuit created between the gate and the channel. This becomes more problematic as the length of the gate in CMOS technology decreases, since the thickness of the gate oxide shrinks (to avoid short channel effects) [16].

2.4.2   Low Transconductance

In general, the transconductance of MOS devices is smaller than that of bipolar devices. The ratio of transconductance g_m to current for a short channel MOS device is

$$\frac{g_m}{I} = \frac{1}{V_{GS} - V_t} = \frac{1}{V_{OV}} \tag{2.9}$$

In order to make up for the low transconductance, the device size needs to be larger. However, this will result in an increased load at the driving state resulting in high power consumption. Another method would be to increase the amplitude of the input signal, which will degrade the linearity of the PA, specifically, the third order non-linearity.

2.4.3   Low f_T and f_{max}

Semiconductor processes often use f_T and f_{max} as a figure of merit to demonstrate the performance of the devices. f_T is the frequency at which the small signal current gain is unity or 0 dB. It is a measure of the maximum frequency the transistor can perform usefully at. It is a function of the technology process and can be defined as

$$f_T = \frac{g_m}{2\pi C_{gs}} \tag{2.10}$$

On the other hand, f_{max} is defined as the frequency where the unilateral gain is a unity of 0 dB. The definition of f_{max} includes the gate resistance that is important in determining the transient response.

$$f_{max} = \frac{f_T}{2\sqrt{g_{ds}(R_g + R_S) + 2\pi f_T R_g C_{gd}}} \tag{2.11}$$

At high RF frequencies, and as the frequency of operation becomes closer to f_T and f_{max}, the performance of the devices degrades, causing low gain and efficiency. Newer technologies such as Silicon-On-Insulator (SOI) and Fully Depleted Silicon-On-Insulator (FDSOI) are exhibiting higher f_T and f_{max} than their bulk CMOS counterparts allowing better performance at higher GHz frequencies. However, the effect and degradation have to still be taken into consideration.

2.5   Cascoding for Larger Power

By stacking transistors, a higher supply voltage can be used. Stacking N devices allows us to use a maximum of $N V_{DD}$. This results in an increased output power while overcoming the breakdown voltage of the device. This also provides better isolation at the input and output stages. The voltage swing across each device is controlled by the bias voltage applied at the gate of the cascode transistor along with a parallel capacitor. The smaller the capacitor, the larger the signal amplitude on the input transistor and vice versa. Usually, the bias voltage of the cascode transistor is higher than that of the input transistor. It can also be connected to V_{DD} resulting in a higher power gain and high linearity.

Cascoding multiple devices has been employed for RF frequencies in FDSOI and SOI devices and has been studied in multiple papers. This is due to the fact that SOI technologies do not suffer from the junction breakdown voltage issues present in bulk CMOS technologies. Some stacking techniques in SOI technologies have been explored in linear PA designs such as in [12, 13] where a popular efficiency enhancement architecture (Doherty PA) in 45 nm SOI was employed. Utilizing multiple transistors has also been investigated in some non-linear switching-like PAs in [17–19]. Most of these studies focus on applying the input signal only to the bottom input transistor in a *stacked-like* topology. Applying a cascoding technique in a switching PA where the switching is applied to all (bottom and top) transistors is yet to be explored more extensively. Such work was presented in [4] where 45 nm SOI was used to implement a multi-output stacked Class-E amplifier at 45 GHz. Each stacked device is driven by a 50% duty cycle signal where the voltage swing is divided between the devices equally.

References

1. H. Wang, K. Sengupta, *RF and mm-Wave Power Generation in Silicon*, 1st edn. (Academic, New York, 2015)
2. Z. Wang, *Envelope Tracking Power Amplifiers for Wireless Communications* (Artech House, Norwood, MA, 2014)
3. A. Afsahi, High power, linear CMOS power amplifier for WLAN applications, Ph.D. dissertation, University of California, 2013
4. A. Chakrabarti, H. Krishnaswamy, Multi-output stacked class-E millimetre-wave power amplifiers in 45 nm silicon-on-insulator metal–oxide–semiconductor: Theory and implementation. IET Microwav. Anten. Propag. **9**(13), 1425–1435 (2015)
5. A. Chakrabarti, H. Krishnaswamy, High power, high efficiency stacked mmwave class-E-like power amplifiers in 45 nm SOI CMOS, in *Proceedings of the IEEE 2012 Custom Integrated Circuits Conference* (2012), pp. 1–4
6. J. Joung, C.K. Ho, K. Adachi, S. Sun, A survey on power-amplifier-centric techniques for spectrum- and energy-efficient wireless communications. IEEE Commun. Surv. Tutor. **17**(1), 315–333 (first quarter 2015)
7. Z. Wang, Demystifying envelope tracking: use for high-efficiency power amplifiers for 4G and beyond. Microw. Mag. IEEE **16**, 106–129 (2015)
8. R.S. Pengelly, The Doherty power amplifier, in *2015 IEEE MTT-S International Microwave Symposium*, May 2015, pp. 1–4
9. N. Elsayed, H. Saleh, A. Mustapha, B. Mohammad and M. Sanduleanu, "A 1:4 Active Power Divider for 5G Phased Array Transmitters in 22nm CMOS FDSOI" 2021 IEEE International Symposium on Circuits and Systems (ISCAS), 2021.
10. T. Instruments, Disentangle RF amplifier specs: output voltage/current and 1 dB compression point (2018) [Online]. Available: https://e2e.ti.com/
11. V. Diddi, H. Gheidi, J. Buckwalter, P. Asbeck, High-power, high-efficiency digital polar Doherty power amplifier for cellular applications in SOI CMOS, in *2016 IEEE Topical Conference on Power Amplifiers for Wireless and Radio Applications (PAWR)*, 2016, pp. 18–20
12. A. Agah, B. Hanafi, H. Dabag, P. Asbeck, L. Larson, J. Buckwalter, A 45 GHz Doherty power amplifier with 23% PAE and 18 dBm output power, in 45 nm SOI CMOS, in *2012 IEEE/MTT-S International Microwave Symposium Digest*, June 2012, pp. 1–3
13. A. Agah, H.T. Dabag, B. Hanafi, P.M. Asbeck, J.F. Buckwalter, L.E. Larson, Active millimeter-wave phase-shift doherty power amplifier in 45-nm SOI CMOS. IEEE J. Solid-State Circ. **48**(10), 2338–2350 (2013)
14. B. Wicks, E. Skafidas, R. Evans, A 60-GHz fully-integrated Doherty power amplifier based on 0.13-um CMOS Process, in *2008 IEEE Radio Frequency Integrated Circuits Symposium*, June 2008, pp. 69–72

15. N. Kalantari, J.F. Buckwalter, A 19.4 dBm, Q-band class-E power amplifier in a 0.12 μm SiGe BiCMOS process. IEEE Microw. Wirel. Compon. Lett. **20**(5), 283–285 (2010)
16. C. Wang, CMOS power amplifiers for wireless communications, Ph.D. dissertation, University of San Diego, 2003
17. S. Pornpromlikit, J. Jeong, C.D. Presti, A. Scuderi, P.M. Asbeck, A 33-dBm 1.9-GHz silicon-on-insulator CMOS stacked-FET power amplifier, in *2009 IEEE MTT-S International Microwave Symposium Digest*, June 2009, pp. 533–536
18. A. Agah, H. Dabag, B. Hanafi, P. Asbeck, L. Larson, J. Buckwalter, A 34% PAE, 18.6 dBm 42–45 GHz stacked power amplifier in 45 nm SOI CMOS, in *2012 IEEE Radio Frequency Integrated Circuits Symposium*, June 2012, pp. 57–60
19. A. Chakrabarti, H. Krishnaswamy, High-power high-efficiency class-E-like stacked mmWave PAs in SOI and bulk CMOS: theory and implementation. IEEE Trans. Microw. Theory Tech. **62**(8), 1686–1704 (2014)

Chapter 3
Doherty Power Amplifier

3.1 Doherty Power Amplifier Design

The Doherty Power Amplifier (DPA) consists of 2 classes of amplifiers in parallel (Class-A and Class-C) connected with a quarter wavelength transmission line combiner:

1. Main (Carrier) amplifier

   - Biased at Class-A/AB
   - Active at low output power

2. Auxiliary (Peaking) amplifier

   - Biased at Class-C
   - Conducts half of the cycle when carrier PA runs into compression

The Doherty amplifier technique is based on the load impedance change of each amplifier, referred to as load modulation, according to the input power level. The design of the PA is divided into three steps: first, the design of the Class-A amplifier, then Class-C, followed by combining the two circuits utilizing quarter wavelength transmission lines. Before discussing the design of the integrated Doherty, transistor sizing, quarter wavelength transmission line design will be discussed.

3.1.1 Transistor Sizing

In order to pick the suitable sizing and bias current for the Class-A amplifier, a post-layout simulation to find the maximum f_T of the transistor is performed. The circuit for this simulation is shown in Fig. 3.1.

To find the maximum f_T at each current density we do the following:

© The Author(s), under exclusive license to Springer Nature Switzerland AG 2022
N. Elsayed et al., *High Efficiency Power Amplifier Design for 28 GHz 5G Transmitters*, Analog Circuits and Signal Processing,
https://doi.org/10.1007/978-3-030-92746-2_3

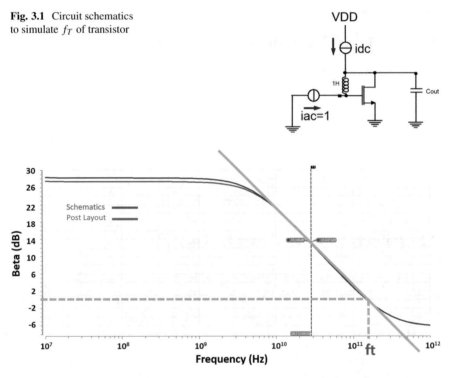

Fig. 3.1 Circuit schematics to simulate f_T of transistor

Fig. 3.2 Extrapolating to find frequency at which $\beta = 0$

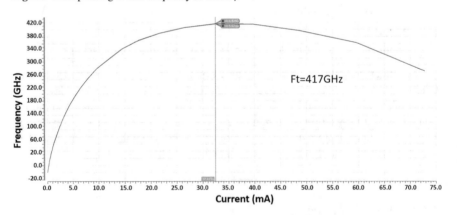

Fig. 3.3 Simulation results of f_T vs. drain current of a 70 μm transistor

1. Calculate current gain (β) for a range of frequencies.
2. Extrapolate to find frequency of which unity gain (f_T) as shown in Fig. 3.2.
3. Sweep current to change the current density for a 70 μ transistor.
4. Plot f_T over different current densities to find ft max (Fig. 3.3).

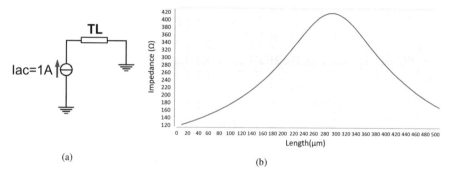

(a)   (b)

Fig. 3.4 (**a**) Simulation of quarter wavelength transmission line. (**b**) Impedance vs. length of transmission line

The maximum f_T simulated was found to be 417 GHz. Since the frequency of operation is at 28 GHz, we do not need to be operating at such a high f_T. Thus, a transistor of 70 μm width and 5 mA bias current was chosen.

3.1.2 Design of Quarter Wavelength Transmission Line

In order to find the quarter wavelength transmission line (TL) at 28 GHz, the schematic level simulation setup in Fig. 3.4a was performed. A transmission line was grounded and connected to an AC current source of magnitude 1 A. At 28 GHz, the length of the TL was varied from minimum length (10 μm) to maximum (1.5 mm), and the impedance was calculated. Since the current is unity magnitude, the impedance is the voltage at the same node. The point at which the impedance is maximum as shown in Fig. 3.4b is the quarter wavelength transmission line length.

3.1.3 Doherty Power Amplifier Integration

The current of the main and auxiliary PAs can be represented as a function of the input to the device. Assuming x to describe the current evolution from DC ($x = 0$) to a maximum current I_M ($x = 1$), for $-\frac{\theta_x}{2} \leq \theta \leq \frac{\theta_x}{2}$, the current is given by

$$i = \frac{I_M}{1 - \cos\left(\frac{\theta_M}{2}\right)} \cdot \left[x * \cos(\theta) - \cos\left(\frac{\theta_M}{2}\right) \right] \qquad (3.1)$$

where θ_M is the value of the current conduction angle for $x = 1$ and corresponds to the maximum current swing I_M. For any fraction x of the input drive, the relationship is described as

$$cos\left(\frac{\theta_M}{2}\right) = x * cos\left(\frac{\theta_x}{2}\right) \tag{3.2}$$

For the main PA, a ratio of the bias voltage and the maximum swing can be defined as

$$\zeta \triangleq \frac{I_{DC,Main}}{I_{M,Main}} \tag{3.3}$$

The conduction angle θ_M of the Class-A PA can now be expressed as θ_A, where:

$$\theta_A = 2\pi cos^{-1} \frac{\zeta}{1-\zeta} \tag{3.4}$$

The bias for the main device can then be described as

$$I_{DC,Main} = -\frac{cos(\frac{\theta_A}{2})}{1 - cos(\frac{\theta_A}{2})} * I_{M,Main} \tag{3.5}$$

The bias current of the auxiliary can be defined similarly relating its conduction angle to the input drive level at which it turns on x_{break} where:

$$I_{DC,Aux} = -\frac{x_{break}}{1 - x_{break}} * I_{M,Aux} \tag{3.6}$$

For low input power levels $0 \leq x \leq x_{break}$, only the main PA is on and the efficiency of the DPA η_{DPA} is equal to the efficiency of the Class-A PA η_A.

For higher input power levels $\leq x_{break}x \leq 1$, the auxiliary PA turns on and the combined efficiency is

$$\begin{cases} P_{out,DPA}(x) = \frac{1}{2} * V_L(x) * [I_{1,Main}(x) + I_{1,Aux}(x)] \\\\ P_{DC,DPA}(x) = P_{DC,Main}(x) + P_{DC,Aux}(x) \\\\ \eta_{DPA} = P_{out,DPA}(x)/P_{DC,DPA}(x) \end{cases} \tag{3.7}$$

Two cascode MOSFET amplifiers are used for the main and auxiliary amplifiers. The lack of the reversed bias diode, between the drain and the source of a transistor in the 22 nm FDSOI process, enables a higher power supply voltage overcoming the low breakdown voltage of bulk CMOS. The technology offers low parasitic capacitances. Consequently, this technology exhibits a high f_T (up to 300 GHz) and has the advantage of a high level of integration. This is very important for mobile applications like 5G. The Class-A amplifier is shown in Fig. 3.5a with the cascoding technique; we can use a power supply up to 2.5 V. The output and input matching circuits were designed to match the amplifier to Z_0=50 Ω. A tapped capacitor output

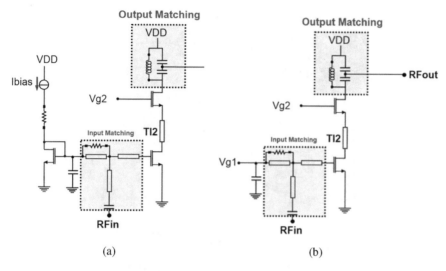

(a)   (b)

Fig. 3.5 Cascode power amplifier circuits. (**a**) Class-A. (**b**) Class-C

matching network was designed to decrease the overall area and reduce the losses, hence increasing the PAE.

The auxiliary amplifier (Class-C) shown in Fig. 3.5b is of the same size as Class-A main amplifier. The biasing, however, is controlled by a gate voltage (Vg1). This, then, allows self-biasing to kick-in at larger input powers turning on more the Class-C auxiliary.

In a balanced Doherty amplifier, the carrier to peak cell size ratio is 1:1. Therefore, a 3-dB splitter (quadrature hybrid) splits the RF signal into two paths with a 90° phase difference. Since we are operating at 28 GHz, this can be achieved using a quarter wavelength transmission line that can easily fit on-chip. In the low power input region, the auxiliary amplifier (Class-C) is off and is seen as an open circuit. This results in a $2Z_0$ load impedance seen by the main amplifier. As the input power level increases, the auxiliary amplifier starts to turn on and the matching network modulates the main amplifier load impedance from $2Z_0$ to Z_0.

The output of the main cell passes through an impedance transformation of Z_0 and a 90° phase shift to compensate for the phase difference introduced at the input. Then it is summed with the output of the auxiliary cell followed by another impedance transformation with an impedance of $Z_0/\sqrt{2}$ and 90° phase shift at the final output to match to 50 Ω as shown in Fig. 3.6. However, we will still need a power divider at the input. This can be implemented at a later stage, on-chip. For this work, we used an external power divider.

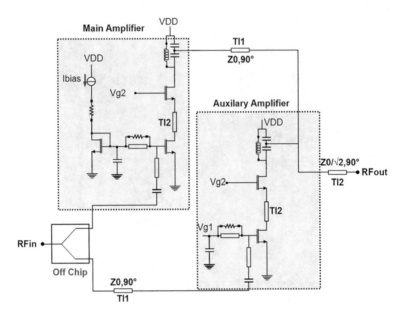

Fig. 3.6 Doherty power amplifier schematics

3.2 Simulation and Measurements

The purpose of this section is to explore the design variables space of the modified Doherty Amplifier and make the best choices for its implementation. Changing the biasing conditions of the auxiliary amplifier changes the performance in efficiency and linearity (1-dB compression point). The biasing of the auxiliary was changed from 100 mV to 500 mV in order to obtain the desired performance. Figure 3.7 shows schematic level, harmonic balance, simulation results, including efficiency, output power, and S-parameters.

In Fig. 3.7a, the output power was swept as a function of input power for a different biasing of the auxiliary. A larger saturated power is obtained for 500 mV auxiliary bias, whereas a smaller biasing will have a larger 1-dB compression point related to the input. Figure 3.7b shows the input and output return losses and the power gain. The input matching ($S_{11} < -10$ dB) is broadband (21 GHz–40 GHz). The output matching is better than -10 dB from 26 GHz up to 32 GHz.

For different biasing levels of the auxiliary, drain efficiency (DE) peaks at different input powers. The maximum DE of 26% is obtained for 100 mV auxiliary bias. From Fig. 3.7d, as the linearity improves, PAE starts degrading. The best trade-off between linearity and efficiency needs to be found. The best performance (where the PA was able to maintain high efficiency and linearity over a wide range of input powers) was achieved when the auxiliary amplifier was biased at 300 mV.

Figure 3.8 shows the photo of the fabricated chip in 22 nm FDSOI CMOS. Highlighted in yellow the main and auxiliary amplifiers and the output combiner.

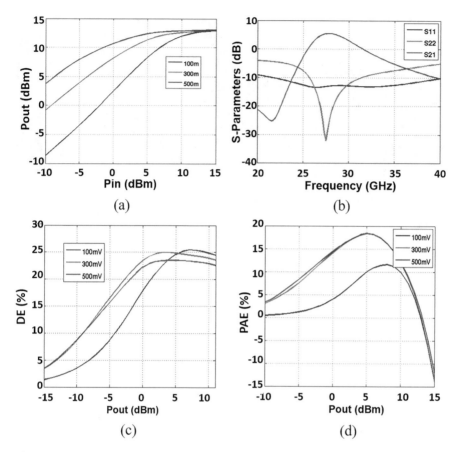

Fig. 3.7 Schematic level, harmonic balance, simulation results. (**a**) Doherty output power vs. input power for a different biasing of the auxiliary PA. (**b**) S-parameters of the Class-A PA. (**c**) Doherty drain efficiency for a different biasing. (**d**) Doherty PAE for a different biasing of the auxiliary PA

The chip occupies 1 mm × 1.2 mm of real estate, on-chip. An external Antristu DC-40 GHz power divider (K240C) with 3-dB loss was used to connect to the two RF inputs related to the main and auxiliary amplifiers.

The measured input return loss shows broadband input matching from 5 GHz up to 29 GHz ($S_{11} < -10$ dBm), whereas the measured output return loss (see Fig. 3.9a) shows small bandwidth matching ($S_{22} < -10$ dBm) around 30 GHz. The measured input return loss was obtained by connecting the RF input only to the main amplifier input. The other input exhibits comparable input return loss. Some discrepancies between the measured and simulated S-parameters were noticed. This is due to the immature PDK models that were used at this point, specifically for RF components.

Figure 3.11 shows the power of the fundamental and its compression and the power of the third order intermodulation along with its compression. From this, we

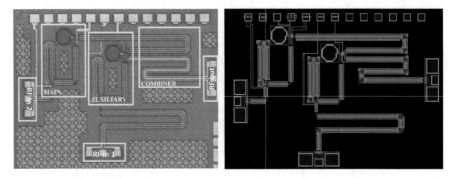

Fig. 3.8 Chip photo and layout (area = 1 mm × 1.2 mm)

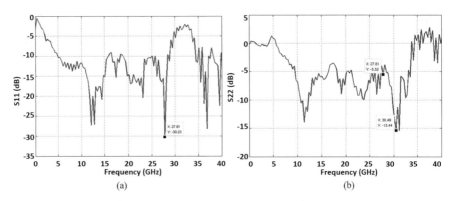

Fig. 3.9 Measured S-parameters of the main PA. (**a**) Measured input return loss (S_{11} parameter). (**b**) Measured output return loss (S_{22} parameter)

observe a 10.5 dB small signal power gain, a 12 dBm compression point (1 dB), a 12.5 dBm saturated power, and 20 dBm IIP3. The power added efficiency PAE is shown in Fig. 3.10. The peak PAE of 16% is obtained at 5 dBm input power, and in 6 dB back-off it retains a value of 12.5%. Table 3.1 shows the benchmark and comparison with other designs from the literature in terms of technology, frequency, peak PAE, 6-db input back-off PAE, and gain. Comparing out work with the existing literature, in [1], a higher peak efficiency was achieved at 42 GHz. However, a comparable PAE at 6-dB back-off was recorded (12%), and a lower power gain of 6 dB. At higher frequencies and mm-Waves (as in [2] and [3]) much lower efficiencies were recorded with lower supply voltages. The design from [4], at 14 GHz, although achieves the highest peak PAE, it has only 8 dB of power gain. Based on the comparison, this design provides an improved trade-off between PAE in back-off and a large power gain with a high linearity and saturated output power (Fig. 3.11).

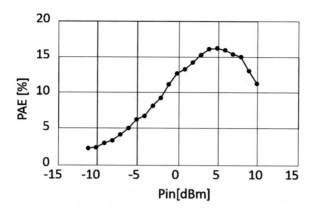

Fig. 3.10 Doherty measured PAE vs. input power

Table 3.1 Benchmark with CMOS Doherty PAs in the literature

Ref.	Tech.	Freq. (GHz)	Supply (V)	Peak PAE (%)	Back-off PAE (%)	Gain (dB)
This work	22 nm CMOS FDSOI	28	2.5	16	12.5	10.5
2013 [1]	45 nm CMOS SOI	42	2.5	21	12	6
2008 [2]	0.13-μm CMOS	60	1.6	3	1.5	7.8
2014 [3]	40 nm CMOS	77	0.9	12	5.7	8.5
2016 [4]	45 nm CMOS SOI	14	2.4	24	20	8

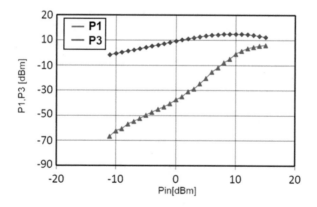

Fig. 3.11 Measurement results of the power of output fundamental (P1), and the power of the third order intermodulation (P3) vs. input power

3.3 Summary

A 28 GHz modified Doherty PA was implemented in the 22nm FDSOI CMOS technology from GF. The PA utilizes two-stack amplifiers as the main and auxiliary amplifiers with 2.5 V power supply. The choice of the technology along with incorporating CPW transmission lines for the splitting and combining circuits maximizes the performance. The measured input return loss shows broadband input matching from 5 GHz up to 29 GHz ($S_{11} < -10$ dB), whereas the measured output return loss shows small bandwidth matching ($S_{22} < -10$ dB) around 30 GHz. The power amplifier achieves peak PAE of 16%, 12.5% at 6-dB back-off and occupies an area of 1.2 mm^2. The measured IIP3 was 20 dBm.

References

1. A. Agah, H.T. Dabag, B. Hanafi, P.M. Asbeck, J.F. Buckwalter, L.E. Larson, Active millimeter-wave phase-shift doherty power amplifier in 45-nm SOI CMOS. IEEE J. Solid-State Circ. **48**(10), 2338–2350 (2013)
2. B. Wicks, E. Skafidas, R. Evans, A 60-GHz fully-integrated Doherty power amplifier based on 0.13-μm CMOS process, in *2008 IEEE Radio Frequency Integrated Circuits Symposium*, June 2008, pp. 69–72
3. E. Kaymaksut, D. Zhao, P. Reynaert, E-band transformer-based Doherty power amplifier in 40 nm CMOS, in *2014 IEEE Radio Frequency Integrated Circuits Symposium*, June 2014, pp. 167–170
4. C.S. Levy, V. Vorapipat, J.F. Buckwalter, A 14-GHz, 22-dBm series Doherty power amplifier in 45-nm CMOS SOI, in *Compound Semiconductor Integrated Circuit Symposium (CSICS), 2016 IEEE* (IEEE, New York, 2016), pp. 1–4

Chapter 4
Delayed Switched Cascode Class-E Amplifier

4.1 Doherty Power Amplifier Design and Architecture

Aside from linear PAs, there has been a growing interest toward non-linear, or switched mode CMOS PAs for mm-Wave applications. These PAs are usually more efficient than the linear counterparts due to their theoretically lossless operation (100% efficiency) [1].

The Class-E PA has been of particular interest due to its simpler output matching network. However, challenges arise due to the lack of ideal square wave signals to drive the PA resulting in soft switching, and low PAE attributed to switching losses [2, 3]. It is also difficult to obtain ideal Class-E efficiency since the operating frequency is a large fraction of the devices' f_T; however, a semi Class-E operation can still be obtained [4].

In order to improve the overall PA PAE and output power, some work in the literature implements switching the cascode device along with the input device to minimize gate resistance. A multi-driven Class-A PA at 90 GHz was implemented in [5] and reported a 4% improvement in PAE and 2 dBm higher output power than a conventional cascode topology.

4.1.1 Classical Class-E

The classical Class-E PA consists of three main design configurations: The input matching, output matching network needed to transform the 50 Ω load impedance into the correct impedance for Class-E operation, and the choke inductor [6]. In an ideal Class-E amplifier, the transistor acts as a switch driven with 50% duty cycle. Traditionally, the Class-E PA utilizes the stray capacitance and an inductor (L_0, C_0)

© The Author(s), under exclusive license to Springer Nature Switzerland AG 2022
N. Elsayed et al., *High Efficiency Power Amplifier Design for 28 GHz 5G Transmitters*, Analog Circuits and Signal Processing,
https://doi.org/10.1007/978-3-030-92746-2_4

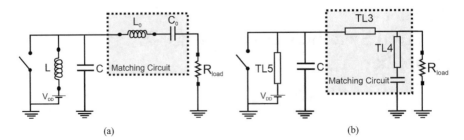

Fig. 4.1 Equivalent circuit of Class-E PA with (**a**) lumped elements matching, (**b**) transmission line matching

for an output network (see Fig. 4.1a). At high frequencies, it is favorable to utilize transmission lines in the design of the matching networks.

The ideal output matching network components of a Class-E PA in Fig. 4.1a are calculated using the equations in [7] to satisfy Zero Voltage Crossing (ZVS) and Zero Current Crossing (ZCS). As a first step, for the parallel-circuit Class-E, the parallel inductance L should be replaced with a short-circuited transmission line TL5 (in Fig. 4.1b) according to

$$Z_0 tan\theta_{TL5} = w_0 L \tag{4.1}$$

where Z_0 and θ_{TL5} are the characteristic impedance and the electrical length of TL5. Using Eq. 4.1, the parallel inductance L can be determined for optimal switched mode operation. The ratio between the transmission line parameters and the load for idealized 50% duty cycle switched mode operation is obtained by

$$tan\theta_{TL5} = 0.732 \frac{R_{Load}}{Z_0} \tag{4.2}$$

Then, the output matching is designed in such a way to transform the optimal load impedance to 50 Ω. A shunt stub and a series transmission line (Tl2, TL3) were designed according to the following equation:

$$Z_L = Z_3 \frac{R_L(Z_4 - Z_3 tan\theta_3 tan\theta_4) + jZ_3 Z_4 tan\theta_3}{Z_3 Z_4 + jR_L(Z_3 tan\theta_4 + Z_4 tan\theta_3)} \tag{4.3}$$

The capacitive stubs provide the needed reactance at the resonant frequency (f_0). The stub lengths are also fixed to have low input impedance at the harmonics to facilitate harmonic suppression [8].

The input matching network is designed as a low pass filter with a peak at 28 GHz. The input impedance of the active device is first determined in simulation with a 50 Ω load. The input network design transforms the impedance at the gate to the 50 Ω load at the source. It consists of a shunt stub to resonate out the input pad capacitance and a series transmission line that combined with the C_{gs} of the input transistor provides a low pass filter. The transmission line input matching network maximizes the voltage swing at the gate, reducing the switching time of the device.

One of the big disadvantages of the Class-E amplifier is that the voltage swing across the device is very large nearly $2V_{DD}$, and hence, a *stacked-like* cascode Class-E amplifier was explored in order to overcome this issue.

4.1.2   Cascode Class-E

In the cascode Class-E topology in Fig. 4.2a, a transistor of the same size is connected in series above the input transistor. The effect of increasing the size of the input transistor on the PAE was simulated. It was found that the PAE is maximum for a device size ratio (W1:W2) between 1:1 to 2:1.

The bigger the bottom transistor gets, the more input driving power it needs [9]. A 1:1 ratio was chosen in order to overcome this problem and satisfy the impedance matching requirements. The intermediate node should be able to sustain Class-E voltage swings. The amplitude should also be scaled in a way that the voltage is shared equally among both devices.

The cascode topology also provides more stability at higher frequencies by improving reverse isolation. In the 22 nm FDSOI technology, the nominal voltage of each device is 0.8 V. To achieve sustained device performance, the maximum swing

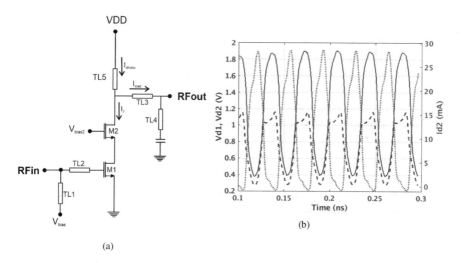

(a)

(b)

Fig. 4.2 (a) Schematic diagram of cascode Class-E PA. (b) Transient simulation of drain current and voltage of the top and bottom devices in cascode Class-E PA. The current and voltage show a non-overlapping pattern exhibiting a Class-E performance. Voltage scaling is also evident from the drain voltages of the input and the cascode device

across each transistor should never increase beyond $2V_{DD}$. The device needs to be large enough to speed up the discharging of the capacitor at the drain of the device and satisfy ZVS conditions. I_{DS} vs. V_{GS} simulations were performed with a range of transistor widths.

The transistor needs to be able to hold the maximum needed drain current while biased at the threshold voltage. Biasing at the threshold voltage provides a compromise of needing a smaller change in the input signal to drive the transistor to peak output current; this results in higher gain but more power dissipation and less efficiency. A transistor of width $70\,\mu$m was picked with the minimum length of the technology.

Figure 4.2b shows the voltage and current swings across the cascode transistor (output transistor) and the voltage swing across the input transistor. It can also be observed from the waveform that the overlap between output voltage and current is minimal, which results in a higher efficiency resembling class-E mode of operation. A tuning shunt inductor can also be added at the intermediate node in order to improve the voltage swing to match a more ideal Class-E operation. However, this technique requires a much bigger area (due to the inductor) that also has a limited quality factor. A series DC-blocking capacitor needs to be added, which has a poor quality factor at mm-Wave frequencies [9].

4.1.3   Switched Cascode Class-E with Tunable Transmission Line

The switched cascode Class-E PA is based on switching not only the input transistor, but also the cascode transistor. It also includes adding a delay between the two transistors in such a way that varying the delay would vary the overlap between the output voltage and the current waveforms and hence vary the efficiency. The schematics of the switched cascode Class-E amplifier are shown in Fig. 4.3. The delay element is added between the input and the cascode transistor and is varied in order to achieve a higher PAE. It is critical to have the transmission line to be tunable for two reasons:

1. The inaccuracy of the PDK would result in variation between the simulated and measured optimal delays.
2. In order to provide flexibility during measurement by adjusting the delay to the optimal delay that results in maximum PAE.

The addition of the tunable transmission line between the two transistors forms a close loop, which might trigger instability and oscillation. This can be an issue at higher carrier frequencies but not for not at 28 GHz with a small delay value. The feedback loop is only through the C_{GD} of the lower transistor that is very small.

The concept of the tunable delay element is shown in Fig. 4.4 The capacitance of the line is affected by the plates connected to C-bit. The inductance of the line is affected by controlling the inductance of the line from L-bit. Switching is

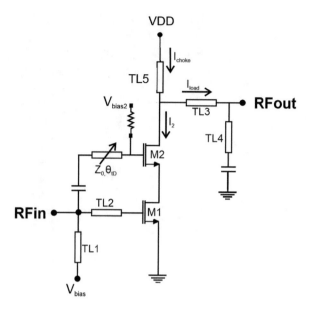

Fig. 4.3 Schematic diagram of switched cascode Class-E PA. The cascode transistor has a switching input with a 50% duty cycle. The phase of the input is controlled using the tunable transmission line with θ_{t_D} delay

Fig. 4.4 Tunable
transmission line concept

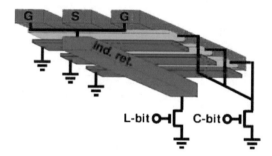

done in such a way that the impedance of the line $Z_0=50\ \Omega$ is constant in all switching conditions as long as the inductance and capacitance controls are changed simultaneously. When the switching transistors connected to the inductance and capacitance are connected to either ground or left open, this changes the inductance and capacitance lines in the same ratio by adjusting the potential difference between these lines to the ground. By this, only the delay $t_d = \sqrt{LC}$ is changed and not the characteristic impedance of the line $Z_0 = \sqrt{L/C}$. The transmission line consists of 7 sections configured in a 4-2-1 structure to allow for a binary weighted control of the phase. The inductance and capacitance lines are driven by transistors as switches that were sized to offer low resistance in the ON state. This work is based on the

Fig. 4.5 Measured maximum (Tline1) and minimum (Tline2) of the group delay of the tunable transmission line

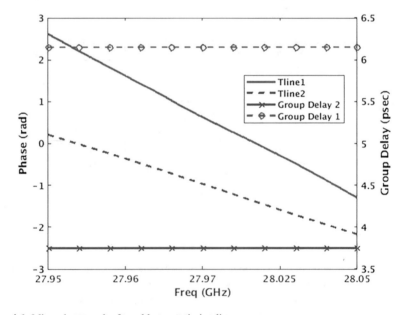

Fig. 4.6 Microphotograph of tunable transmission line

same concept of a variable delay transmission line presented in [10] and [11] but implemented in 22 nm FDSOI.

The chip microphotograph of the tunable T-Line is shown in Fig. 4.6. The signal layer was implemented in the top (LB) layer: followed by the capacitance grid on the 2nd most top layer (OI) followed by the inductance line layer (C6). Measurements show a linear-phase behavior (see Fig. 4.5) and constant group delay in two extreme

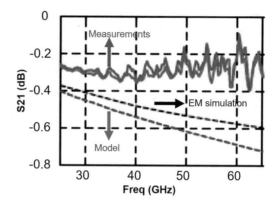

Fig. 4.7 Measured and simulated S_{21} magnitudes of tunable transmission line

conditions. As shown in Fig. 4.5, the delay of the line can vary between 3.7 ps and 6.2 ps with varying settings. The S_{21} parameter of the tunable transmission line was measured against frequency. At 28 GHz, the losses are very small (0.3 dB) as shown in Fig. 4.7, which would not have much effect on the switching mechanism. The precision of the transmission line as a passive component is very high due to lithography precision. The length of the implemented transmission line was chosen based on the simulated range of delays needed.

For a higher range, a longer line needs to be implemented, and the change in losses needs to be accounted for. For simple analysis, the output and input devices are represented as lossless switches with an output capacitance of $C_{o,(1,2)}$ and an ON-resistance of $R_{ON,(n)}$, where n denotes the nth transistor in a stack. The output capacitance $C_{o,2}$ can be represented as

$$C_{o,2} = C_{d,2} + C_{gd,2} \qquad (4.4)$$

where $C_{d,2}$ is the drain-to-ground capacitance of the cascode device, and $C_{gd,2}$ is the gate-to-drain capacitance of the same device.

For efficiency analysis, we can define the currents through the output transistor. Having a delay between the gate of the cascode transistor and the input transistor results in a delay when the current in the cascode transistor peaks. Therefore, a logical way of modelling is to delay the current in the cascode transistor [9, 12]. Hence, the load network current during the ON cycle of the transistor can be defined as

$$I_{load} = i_2 cos(w_0 t + \theta_2 + \theta_{t_D}) \qquad (4.5)$$

where i_2 is the load network current. Hence, the current through the output switch can be defined as

$$I_2 = I_{choke} - I_{load} \qquad (4.6)$$

where I_{choke} is the current through the choke inductor. The power loss through the switch can then be defined as

$$P_{loss} = I_{RMS,2}^2 * 2R_{ON} \tag{4.7}$$

where R_{ON} is the output switch ON-resistance assuming both devices are of identical size. From Eqs. 4.6 and 4.7, we can define and PAE as

$$
\begin{aligned}
DE &= 1 - \frac{P_{loss}}{P_{DC}} \\
&= 1 - \frac{I_{choke} - i_2 cos(w_0 t + \theta_2 + \theta_{t_D}) * 2R_{ON}}{V_{DD} * I_{DC}} \\
PAE &= 1 - \frac{P_{loss}}{P_{DC}} - \frac{P_{in}}{P_{DC}} = DE - \frac{P_{in}}{V_{DD} * I_{DC}}
\end{aligned}
\tag{4.8}
$$

From Eq. 4.8 it may seem that by minimizing the ON-resistance and increasing the device size, DE will increase. However, if the device gets too large, then the input drive will need to increase, which will have a negative effect on the PAE. Therefore, the device size needs to be chosen to provide a trade-off between the PAE and DE. This way, a new term (θ_{t_D}) is introduced to provide more control to achieve the best efficiency for the PA. Figure 4.9 shows the drain and current waveforms of the PA, with varying (θ_{t_D}).

From Fig. 4.9, the overlap between the drain current and the voltage varies with the varying time delays; hence, one can achieve maximum PAE by adjusting the delay to result in minimum overlap and PAE as depicted in Fig. 4.8. The results also show an improvement in PAE by 8% through applying the switched cascode mechanism compared to a constant bias cascode topology (Fig. 4.8).

4.2   Implementation and Measurement Results

The implemented PA in Fig. 4.10 was designed and fabricated in 22 nm FDSOI from GlobalFoundries and was tested using on-chip probing using the Elite 300 semi-automatic probe station. The setup was calibrated up to the probe tips using the RHODE & SCHWARZ ZVB8 Vector Network Analyzer (VNA). The input signals were generated using the RHODE & SCHWARZ SMF 100A signal generator. All simulations were performed in a post-layout RC extracted environment utilizing load-pull and harmonic balance.

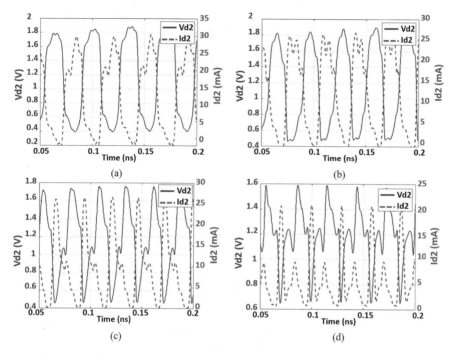

Fig. 4.8 Varying PAE with varying θ_{t_D}. Changing the phase of the cascode transistor input affects the PAE of the PA. Maximum efficiency was achieved without affecting the output power. The figure shows an improvement in PAE by applying the switching technique compared to a constant biased cascode transistor

4.2.1   Small Signal Measurements

The small signal measurements were done using a VNA. Figures 4.11, 4.12, and 4.13 depict the measured and simulated small signal S-parameters of the cascode Class-E PA. In measurement, a peak gain of 8.5 dBm was achieved at ≈ 28 GHz. A frequency shift of ≈ 1 GHz was observed in both input and output matching parameters (S_{11} and S_{22}) due to mismatch in the estimation of the parasitic capacitance during the simulation phase.

4.2.2   Large Signal Measurements

Two sets of measurements were performed at 28 GHz for the large signal measurements using a spectrum analyzer from RHODE & SCHWARZ. The first one was for the cascode Class-E PA, achieving a peak PAE of 28% and 41% DE. Figure 4.14 shows the measurement versus simulation results. The switched cascode topology achieved better efficiency since the DC power dissipated is reduced, resulting in a

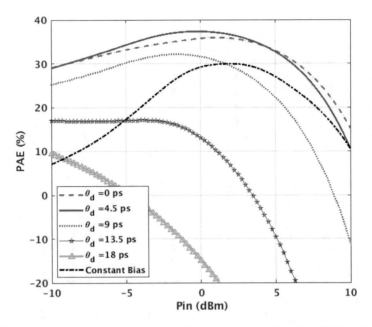

Fig. 4.9 Drain current and voltage waveforms with varying θ_{tD}: (**a**) $\theta_{tD}=0$ ps, (**b**) $\theta_{tD}=4.5$ ps, (**c**) $\theta_{tD}=9$ ps, (**d**) $\theta_{tD}=13.5$ ps. The change in the overlap between the two signals affects the PAE of the PA

peak PAE of 35% and a DE of 45% in Fig. 4.16. The output power and power gain of the PAs shown in Fig. 4.15 remain the same and show a gain of 8.5 dB.

4.2.3   Comparison with the State-of-the-Art Class-E and Cascode PAs

Table 4.1 shows the comparison of both the cascode and switched cascode Class-E PAs to the existing state-of-the-art mm-Wave PAs. To our knowledge, the best PAE efficiency reported for a Class-E at mm-Wave frequency was reported in [2] implemented in 45 nm SOI at 47 GHz. Our work presents a higher peak PAE, along with a higher DE (η). From Table 4.1 we are able to report the best drain efficiency and overall performance for a fully integrated PA at 28 GHz along with the best FoM defined in Eq. 2.7. For 22 nm FDSOI nFET the reported f_T/f_{max} is 347/371 GHz [16].

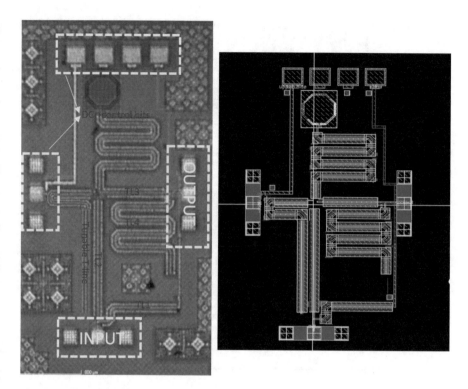

Fig. 4.10 22 nm FDSOI chip microphotograph and layout. The total die size is $1 \times 0.6 \text{ mm}^2$

4.3 Summary

This chapter has presented the novel design and measurement of two cascode based Class-E PAs in 22nm FDSOI CMOS. The first one is based on the cascode topology achieving a peak PAE of 28%. The second one is based on a delayed switched cascode topology using a tunable transmission line for a delay element to achieve minimum overlap between the current and voltage waveforms at the drain of the output device. The switched cascode Class-E PA achieves a PAE of 35% and a peak DE of 45% with a saturated output power of 13 dBm indicating an improvement by 4% and 7% in DE and PAE, respectively, from the classical cascode Class-E topology.

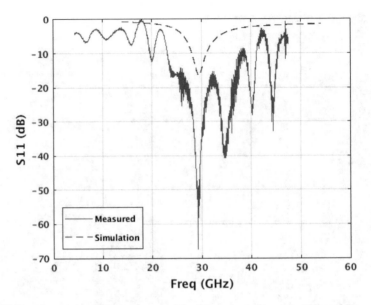

Fig. 4.11 Post-layout simulation vs. measured S_{11}. The measured input matching achieves < -40 dB at 28 GHz

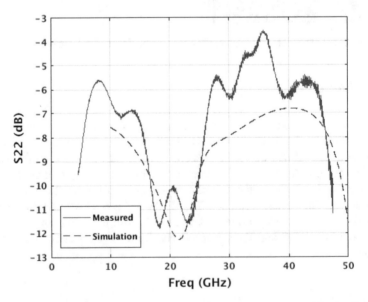

Fig. 4.12 Post-layout simulation vs. measured S_{22}. The measured output matching is -10 dB at 28 GHz

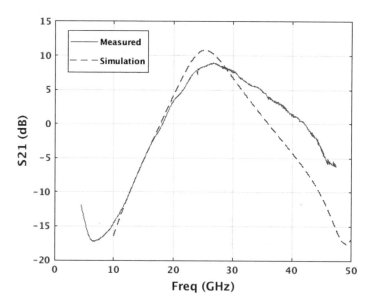

Fig. 4.13 Post-layout simulation vs. measured S_{21}. The measured S_{21} is 8.5 dB with $\approx$1 GHz frequency shift from the simulation

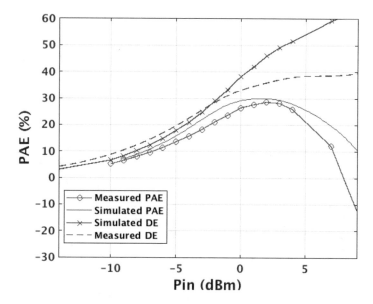

Fig. 4.14 Measured and simulated PAEs, and DE for cascode Class-E PA. The maximum measured DE/PAE for the PA is 41/28%. Peak PAE is achieved at 2 dBm P_{in}

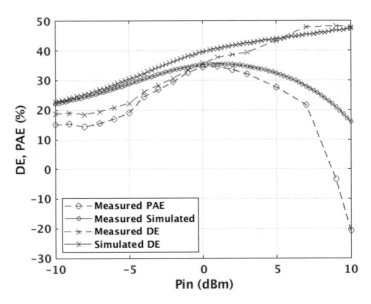

Fig. 4.15 Gain vs. output power for cascode Class-E PA. Measured gain in 8.5 dB with a saturated output power P_{sat} of 13 dBm

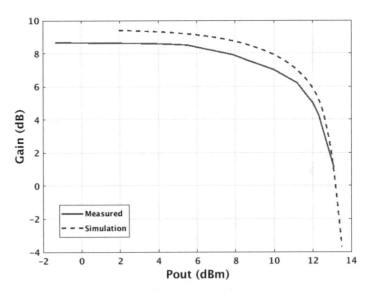

Fig. 4.16 Measured and simulated PAEs, and DE of switched cascode Class-E PA. The maximum measured DE/PAE is 45/35%. Peak PAE is achieved at 1 dBm P_{in}

Table 4.1 Comparison with the state of the art

Ref.	Technology	Freq. (GHz)	Psat (dBm)	η (%)	Peak PAE (%)	Gain (dB)	f_{max} (GHz)	FoM[a]	Class of operation
This work	22 nm FDSOI	28	13	41	28	8.5	370	13.5	Class-E cascode
This work	22 nm FDSOI	28	13	45	35[b]	8.5	370	14.5[b]	Class-E switched cascode
2015 [9]	45 nm SOI	47.5	19.1	24.5	16	8.2	130	10.6	Class-E 2-stacked
2012 [2]	45 nm SOI	47	17.6	42.4	34.6	13	190	13.85	Class-E 2-stacked
2012 [2]	45 nm SOI	47.5	20.3[b]	23	19.4	12.8[b]	180	13.46	Class-E 4-stacked
2019 [13]	40 nm CMOS	2.5	17.5	54[b]	34	N/A	250	N/A	Class-E Doherty
2012 [14]	45 nm SOI	45	18.6	N/A	34	9.5	190	10.9	Class-AB 2-stacked
2010 [15]	0.13 µm SiGe	42	19.4	N/A	14.4	6	240	1.84	Class-E
2012 [4]	32 nm SOI	60	12.5	N/A	30	10	N/A	N/A	Class-E

[a] FoM= P_{sat} (dBm)+Gain(dB)+10log$_{10}$(PAE)+20log$_{10}$ f_0/f_{max}
[b] Highest value per column for each performance measure

References

1. A. Chakrabarti, H. Krishnaswamy, Design considerations for stacked class-E-like MmWave high-speed power DACs in CMOS, in *2013 IEEE MTT-S International Microwave Symposium Digest (MTT)*, June 2013, pp. 1–4
2. A. Chakrabarti, H. Krishnaswamy, high power, high efficiency stacked mmWave class-E-like power amplifiers in 45 nm SOI CMOS, in *Proceedings of the IEEE 2012 Custom Integrated Circuits Conference*, 2012, pp. 1–4
3. O. El-Aassar, M. El-Nozahi, H.F. Ragai, Loss mechanisms and optimum design methodology for efficient mm-Waves class-E PAs. IEEE Trans. Circ. Syst. I: Regul. Pap. **63**(6), 773–784 (2016)
4. O.T. Ogunnika, A. Valdes-Garcia, A 60 GHz Class-E tuned power amplifier with PAE >25 in 32 nm SOI CMOS, in *2012 IEEE Radio Frequency Integrated Circuits Symposium*, June 2012, pp. 65–68
5. A. Agah, J.A. Jayamon, P.M. Asbeck, L.E. Larson, J.F. Buckwalter, Multi-drive stacked-FET power amplifiers at 90 GHz in 45 nm SOI CMOS. IEEE J. Solid-State Circ. **49**(5), 1148–1157 (2014)
6. Z. Xu, E. El-Masry, Design and optimization of CMOS Class-E power amplifier, in *2003 IEEE Int. Symp. Circuits and Systems Dig.*, vol. 1 (2003), pp. I–325
7. N.O. Sokal, A.D. Sokal, Class E- a new class of high-efficiency tuned single-ended switching power amplifiers. IEEE J. Solid-State Circ. **10**(3), 168–176 (1975)
8. A.V. Grebennikov, H. Jaeger, Class E with parallel circuit - a new challenge for high-efficiency RF and microwave power amplifiers, in *2002 IEEE MTT-S International Microwave Symposium Digest (Cat. No.02CH37278)*, vol. 3, 2002, pp. 1627–1630
9. A. Chakrabarti, H. Krishnaswamy, Multi-output stacked class-E millimetre-wave power amplifiers in 45 nm silicon-on-insulator metal–oxide–semiconductor: theory and implementation. IET Microw. Antennas Propag. **9**(13), 1425–1435 (2015)
10. S. Shlafman, B. Sheinman, D. Elad, A. Valdes-Garcia, M.A.T. Sanduleanu, Variable delay transmission lines in advanced CMOS SOI technology, in *2014 IEEE Radio Frequency Integrated Circuits Symposium*, 2014, pp. 111–114
11. W.H. Woods, A. Valdes-Garcia, H. Ding, J. Rascoe, CMOS millimeter wave phase shifter based on tunable transmission lines, in *Proceedings of the IEEE 2013 Custom Integrated Circuits Conference*, 2013, pp. 1–4
12. D. Sira, T. Larsen, Modeling of cascode modulated power amplifiers, in *2011 NORCHIP*, 2011, pp. 1–4
13. M. Hashemi, L. Zhou, Y. Shen, L.C.N. de Vreede, A highly linear wideband polar Class-E CMOS digital Doherty power amplifier. IEEE Trans. Microw. Theory Tech. **67**(10), 4232–4245 (2019)
14. A. Agah, H. Dabag, B. Hanafi, P. Asbeck, L. Larson, J. Buckwalter, A 34% PAE, 18.6 dBm 42–45 GHz stacked power amplifier in 45 nm SOI CMOS, in *2012 IEEE Radio Frequency Integrated Circuits Symposium*, June 2012, pp. 57–60
15. N. Kalantari, J.F. Buckwalter, A 19.4 dBm, Q-band class-E power amplifier in a 0.12 μm SiGe BiCMOS process. IEEE Microw. Wirel. Compon. Lett. **20**(5), 283–285 (2010)
16. J. Hoentschel, L. Pirro, R. Carter, M. Horstmann, 22FDX technologies for ultra-low power IoT, RF and mmWave applications. Compos. Nanoélectron. **2**, 01 (2019)

Chapter 5
Delayed Switched Cascode Doherty Class-E PA

5.1 Proposed Class-E Doherty PA Design

Different techniques have been explored in the literature to improve the Doherty PA performance. These techniques include the stacking technique explored in [1, 2], multi-stage Doherty [3], and inverted and series configurations [4, 5].

However, these techniques are still limited to the maximum efficiency that could be achieved with the Class-C PA [6]. Therefore, switching amplifiers have been explored due to their theoretically high efficiency. One of the most popular switching amplifiers for such techniques is the Class-E PA due to its relatively simple matching network. Utilizing Class-E in a Doherty configuration has been discussed previously in the literature in [6–8] utilizing a classical Class-E configuration.

In this chapter, a high efficiency Class-E PA was utilized instead, which can theoretically achieve 100% efficiency, thus improving the overall PAE at both peak and back-off power levels and resulting in a smoother transition between the two. The design of the cascode Class-E PA (discussed in Chap. 4 was modified to achieve maximum efficiency and performance at peak. The cascode transistor can operate in two modes:

1. **Constant Bias Mode**: The cascode transistor is set at a constant bias to ensure equal division of the voltage swing in both transistors.
2. **Switched Mode**: The cascode transistor is also switched via a tunable transmission line at the gate of the cascode transistor.

By modifying the delay of the tunable transmission line in such a way that minimizes the drain voltage and current waveforms overlap, the efficiency improves further.

Figure 5.1 depicts the architecture of the proposed DPA utilizing a Class-AB PA as a main amplifier and a switched cascode Class-E as an auxiliary amplifier. The

© The Author(s), under exclusive license to Springer Nature Switzerland AG 2022   47
N. Elsayed et al., *High Efficiency Power Amplifier Design for 28 GHz 5G Transmitters*, Analog Circuits and Signal Processing,
https://doi.org/10.1007/978-3-030-92746-2_5

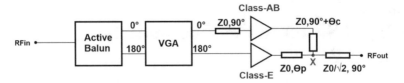

Fig. 5.1 Proposed DPA architecture

auxiliary PA operates as a switch under large signal conditions with minimal power
dissipation due to the non-overlapping characteristics of the current and voltage
waveforms. The input signal is divided equally by an active balun that creates a 180
° out-of-phase signal in order to avoid having an off-chip splitter. Signal levels are
then controlled by a variable gain amplifier (VGA) and are transferred to the main
and auxiliary PAs. A $\lambda/4$ transmission line with $Z_0 = 50\ \Omega$ impedance is placed
at the input of the main PA, and a similar one is placed after the output matching
network of the main PA to provide phase balance between the two paths. At the
output, an impedance transformation in the form of another $\lambda/4$ transmission line
with $\approx 35.4\Omega$ impedance is added. θ_c and θ_p are used for phase compensation.

5.1.1 Active Balun

The active balun is shown in Fig. 5.2. Given that M4 and M5 have the same bias
voltage at the gate, the following equation is true:

$$V_{GS4} + V_{GS3} = V_{BIAS} \tag{5.1}$$

This ensures a Class-AB behavior. Therefore, when the current in M4 increases, the
current in M3 decreases. By choosing correctly the values of R4 resistors, we can
get 180° phase difference between the currents and a differential output at RFout
outputs. The L3∥C6 tank is resonant at the RF (28 GHz). The capacitors C7 and C8
help in getting differential outputs at the resonant frequency of 28 GHz. Figure 5.3
shows the magnitude and phase difference in the two out-of-phase outputs. The
magnitudes of the 2 out-of-phase outputs are identical ensuring the same signal
levels delivered to both the auxiliary and main PAs. Figure 5.4 shows the simulated
output power of the active balun with an 8.2 1-dB compression point.

5.1.2 VGA

The VGA is presented in Fig. 5.5. The differential outputs of the active balun are
applied at the In+ and In- inputs. The currents of the differential pair M2 are

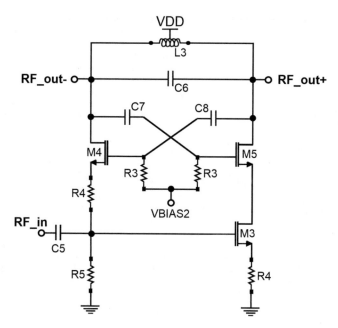

Fig. 5.2 Circuit schematics of the active balun

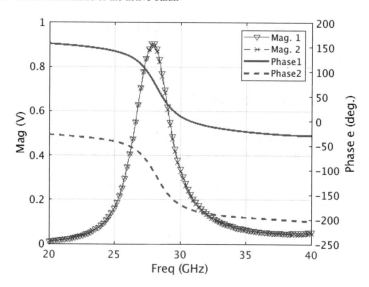

Fig. 5.3 Measured phase and magnitude of the active balun output. Measurement results show equal magnitude of both outputs and a phase difference of 180° at 28 GHz

steered to the differential pair M3. The bias voltage at VG+ and VG- generated from currents I1 and I2 controls the current steering to the RFout nodes. In consequence,

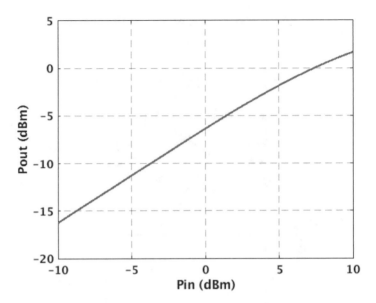

Fig. 5.4 Measured single ended P_{out} vs. P_{in} of the active balun showing an input inferred 1-dB compression point of 8.2 dBm

we get larger or smaller gain accordingly. The L1∥C1 tank is resonant at RF (28 GHz).

5.1.3 Integration of Class-E Doherty PA

When the input signal is lower than the threshold level of the auxiliary PA, the Class-E PA is off, while the main PA is operating. This means that the impedance seen from point **X** in Fig. 5.6 should be infinite. This is ensured by adding a transmission line after the output matching network of the auxiliary PA with Z_0 impedance with an optimal delay of θ_p satisfying the equation:

$$Z_0 \frac{\frac{1}{jw_0C_0} + jZ_0 tan\theta_p}{Z_0 + \frac{tan\theta_p}{w_0C_0}} = 0 \tag{5.2}$$

where C_0 is the parallel capacitance of the Class-E PA. From Eq. 5.2, θ_p is calculated as follows:

$$\theta_p = tan^{-1} \frac{1}{Z_0 w_0 C_0} \tag{5.3}$$

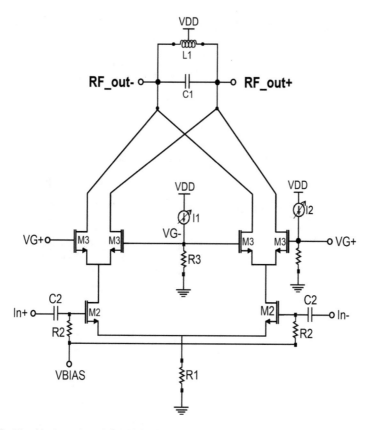

Fig. 5.5 Circuit schematics of the VGA

In the low power region, the Doherty PA behaves as a typical Class-AB PA up reaching its largest output voltage swing at the fundamental $V_{1,Main} = V_{DD} - V_k$ at which the output current reaches a predetermined level I_{break} at point P_{break}. The impedance at the main device up to the break condition is expressed by

$$R_{Main} = \frac{V_{1,Main}}{I_{1,Main}} = \frac{Z_0^2}{R_{Load}} Pin < P_{break} \qquad (5.4)$$

where $V_{1,Main}$ and $I_{1,Main}$ are the voltage and current fundamental components of the main PA, respectively. The ratio of the current at the breaking point I_{break} and the maximum output current of the main PA $I_{M,Main}$ can then be expressed by

$$\frac{I_{break}}{I_{M,Main}} = \frac{P_{break} - \cos\frac{\theta_{AB}}{2}}{1 - \cos\frac{\theta_{AB}}{2}} \qquad (5.5)$$

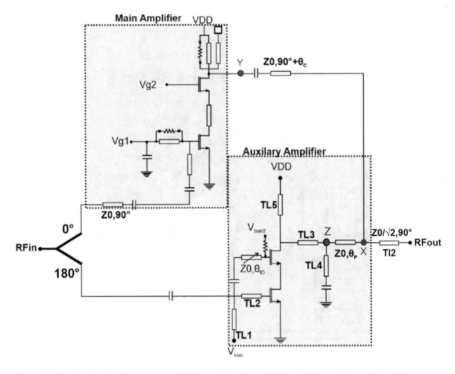

Fig. 5.6 Basic circuit of the proposed DPA with switched Class-E PA as the auxiliary PA

where θ_{AB} is the conduction angle of Class-AB PA. As the input signal increases, and we enter the Doherty region, we can assume that the main PA has a constant voltage source of the maximum voltage swing it can achieve $V_{1,Main}$ and the load impedance at each PA is then defined as

$$R_{Main} = \frac{V_{DD} - V_k}{I_{1,Main}} \tag{5.6}$$

$$R_{Aux} = \frac{V_{DD} - V_k}{I_{1,Aux}} \tag{5.7}$$

The output voltage can then be expressed as

$$V_{Load} = R_{Load} * (I_{1,Main} + I_{1,Aux})$$

(5.8)

$$= R_{Load} * I_{1,Main} \left[1 + \frac{I_{1,Aux}}{I_{1,Main}} \right]$$

For this to be true, two conditions should be satisfied:

1. The fundamental output of both PAs should be in phase at point X. This is achieved by adjusting the phase compensation lines θ_p and θ_c.
2. The maximum power transfer to load from the Class-E satisfying:

$$\theta_Y + \theta_{main} = \theta_Z + \theta_{Aux}$$

(5.9)

where θ_{Main} and θ_{Aux} are the phases of the transmission function of the load network for the main and auxiliary PAs, respectively. θ_Y and $\theta_Y = Z$ are the phases of the fundamental component at points Y and Z in Fig. 5.6. After the design of the output network of the auxiliary PA, θ_{Aux} is calculated from Eq. 5.9 and the load network of the main PA needs to be optimized accordingly. The drain efficiency (η) of the Class-E Doherty PA can then be expressed as

$$\eta = \frac{P_{out}}{P_{DC}} = \frac{P_{out,Main} + P_{out,Aux}}{(P_{DC,Main} + P_{DC,Aux})}$$

(5.10)

$$= \frac{\eta_E P_{DC,E} + \eta_{AB} P_{DC,AB}}{P_{DC,E} + P_{DC,AB}}$$

5.2 Simulation and Measurement

The chip in Fig. 5.7 was implemented in Global Foundries' 22nm FDSOI. The testing was done on-chip using the Elite 300 semi-automatic probe station. The input and output signals are measured using ground-signal-ground (GSG) probes, while the DC signals are through a multi-wedged probe. All simulation results were performed with post-layout, RC extracted harmonic balance simulation (Fig. 5.8).

5.2.1 Small Signal Parameters

A vector network analyzer (VNA) was used to measure the small signal parameters of the PA shown in Figs. 5.9 and 5.10. The VNA was calibrated up to the probe tips. Figure 5.10 shows the simulated vs. measured peak gain of the PA with different

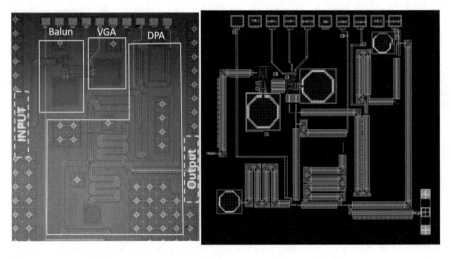

Fig. 5.7 Class-E DPA chip microphotograph and layout ($1.3 \times 1.68\,\text{mm}^2$)

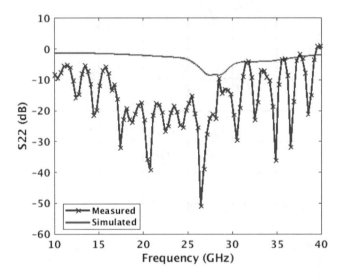

Fig. 5.8 Simulated vs. measured S_{22}. The output matching is below -10 dB at 28 GHz with a frequency shift in measurement

VGA settings. Figure 4.13 shows that the output gain of the PA can be controlled by varying the control inputs of the VGA. The peak measured gain with maximum VGA gain is 17 dB at 28 GHz with an output saturated power of 17.5 dBm. The input matching in Fig. 5.9 shows the broadband input matching of the active balun. The output matching in Fig. 5.8 is well below -10 dB. A frequency shift of around 2 GHz is observed on all S-parameters due to unaccounted parasitic capacitances.

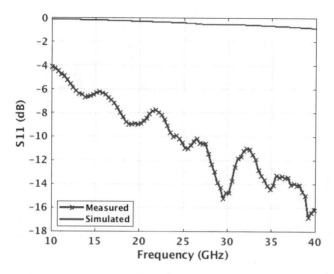

Fig. 5.9 Post-layout simulation vs. measured S_{11}. The input matching of the active balun shows a broadband input matching behavior

There are magnitude deviations between the simulation and measurement results due to discrepancies in parasitic estimation and immature PDK models.

5.2.2 Large Signal Measurements

A spectrum analyzer was used to measure input/output power of the PA under various inputs. Figure 5.11 shows the power gain of the DPA with a saturated output power (P_{sat}) of 17.5 dBm. Two sets of measurements were recorded for efficiency measurements: the first one using the Class-E PA in the constant bias mode (Fig. 5.12) and the second one using the Class-E PA in the switched mode (Fig. 5.13).

The constant bias DPA measured a peak PAE of 28 and 25% at back-off with only a 3% loss. The maximum DE measured was 48%. The switched mode DPA recorded an improvement in both peak and backed-off PAE at 32/31%, respectively. The maximum DE measured was 59%. The measurements also show an improvement in the overall peak and back-off PAE from the Class-C based DPA [9].

Table 5.1 Comparison with the state-of-the-art Doherty CMOS PAs

Ref.	Tech.	Freq. (GHz)	Psat (dBm)	Peak PAE(%)	BO PAE (%)	Gain (dB)	FoM[a]	FoM[a] (BO)	Matching network	Architecture
This work	22nm FDSOI	28	17.5	28	25	17	**29.3**	**28.7**	On-chip	Doherty Class-E
This work	22nm FDSOI	28	17.5	32	31	17	**29.7**	**29.5**	On-chip	Doherty Class-E switched mode
2020 [10]	22nm FDSOI	28	22.5	28.5	22.1	26.1	29.5	28	On-chip	Doherty 3-stacked
2019 [7]	40nm CMOS	2.5	17.5	34	25	29[b]/24[c]	25.9	19.6	Off-chip	Doherty Class-E, digital control
2012 [1]	45nm SOI	42	18	23	17	7	19.4	18.1	On-chip	Doherty
2014 [11]	40nm CMOS	77	16.2	12	5.7	9[b]	25.7	22.5	On-chip	Doherty transformer-based
2008 [12]	0.13um CMOS	60	7.8	3	1.5	13.5	19.3	16.3	On-Chip	Doherty
2016 [5]	45nm SOI	14	22	24	20	8	15.1	14.3	On-Chip	Series Doherty
2011 [13]	90nm CMOS	71–76	11.7	30.6	15.6	4.7	22.8	19.9	On-chip	Doherty
2018 [14]	45nm SOI	28	22.4	40	28	10	26.5	24.9	On-chip	Doherty

Bolded text represents best performance in literature

[a] FoM=P_{sat}(dBm)+Gain(dB)+10log$_{10}$(Peak/BO PAE)+20log$_{10}$ f_0/f_{max}

[b] Estimated graphically

[c] At 6-dB back-off

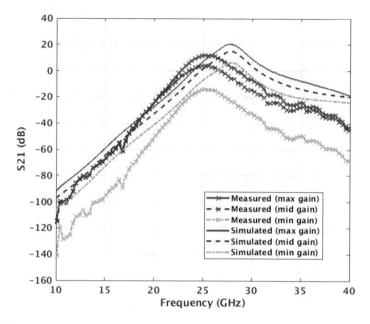

Fig. 5.10 Post-layout simulation vs. measured S_{21} of the whole chain (Balun+VGA+PA). Maximum measured gain of 17 dB. Changing the gain of the VGA results in a change in the maximum gain achieved by the DPA

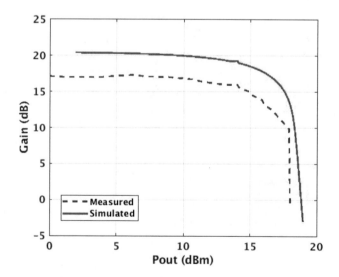

Fig. 5.11 Post-layout simulation vs. measured power gain. Maximum power gain of 17 dB and P_{sat} of 17.5 dBm

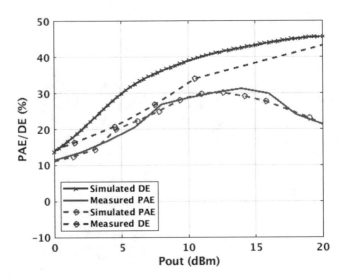

Fig. 5.12 Measured and simulated PAE/DE of constant bias Class-E DPA. The measurements show a peak PAE of 28% with a 3% loss at 6-dB back-off

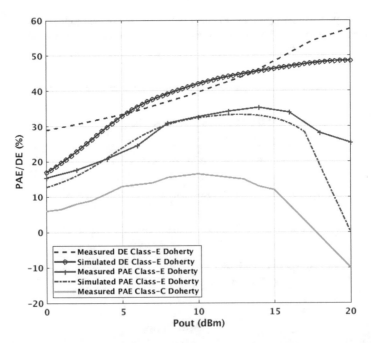

Fig. 5.13 Measured and simulated PAE/DE of switched mode Class-E DPA vs. measured PAE of Class-C DPA. The measured DPA shows a 32% peak PAE with a 1% loss at back-off compared to a 16% peak PAE for classical DPA measurement

5.2.3 Comparison with State of the Art

Table 5.1 shows the comparison of each mode of the DPA to the existing state-of-the-art CMOS DPAs based on Class-C and Class-E as the auxiliary PA. The results for both Class-E DPA modes show significant improvement in minimizing the drop in PAE between the peak and back-off region while maintaining high output power as well as maintaining a high overall PAE. Having the Class-E PA in a delayed switched mode improves the peak PAE by 5% compared to keeping it at a constant bias. Also, the drop in efficiency from peak to the back-off region decreases from 3% to only 1%. The figure of merit (FoM) defined by the ITRS provides a comparison linking the output power, gain, efficiency, and operating frequency of a PA. This is used as a standard to compare the PAs in Table 5.1.

$$
\begin{aligned}
\text{FoM} = {} & P_{sat}(\text{dBm}) + \text{Gain}(\text{dB}) + 10 \log_{10}(\text{PAE}) \\
& + 20 \log_{10} f_0/f_{max}
\end{aligned}
\tag{5.11}
$$

5.3 Summary

In this chapter, a 28 GHz Class-E based Doherty PA was implemented in 22nm FDSOI. The PA utilizes an on-chip active balun followed by a VGA. The stacked topology is used for both the main and auxiliary amplifiers to increase efficiency and output power with a 17 dB gain and 17.5 dBm saturated output power. The auxiliary amplifier (Class-E) is used in two modes: constant bias mode and switched mode. The former had a measured peak PAE of 28 and 25% at 6-dB back-off. The switched mode DPA has a measured peak PAE of 32% with and 31% at 6-dB back-off. This shows a significant improvement in the recorded PAE at both peak and back-off and the best FoM when compared to other DPAs from the literature.

References

1. A. Agah, B. Hanafi, H. Dabag, P. Asbeck, L. Larson, J. Buckwalter, A 45GHz Doherty power amplifier with 23% PAE and 18dBm output power, in 45nm SOI CMOS, in *2012 IEEE/MTT-S International Microwave Symposium Digest* (2012), pp. 1–3
2. A. Agah, H.T. Dabag, B. Hanafi, P.M. Asbeck, J.F. Buckwalter, L.E. Larson, Active millimeter-wave phase-shift Doherty power amplifier in 45-nm SOI CMOS. IEEE J. Solid-State Circuits **48**(10), 2338–2350 (2013)
3. N. Srirattana, A. Raghavan, D. Heo, P.E. Allen, J. Laskar, Analysis and design of a high-efficiency multistage Doherty power amplifier for wireless communications. IEEE Trans. Microw. Theory Tech. **53**(3), 852–860 (2005)
4. M. Lee, S. Kam, Y. Lee, Y. Jeong, Design of highly efficient three-stage inverted Doherty power amplifier. IEEE Microw. Wireless Compon. Lett. **21**(7), 383–385 (2011)
5. C.S. Levy, V. Vorapipat, J.F. Buckwalter, A 14-GHz, 22-dBm series Doherty power amplifier in 45-nm CMOS SOI, in *2016 IEEE Compound Semiconductor Integrated Circuit Symposium (CSICS)* (IEEE, Piscataway, 2016), pp. 1–4

6. C. You, X. Zhu, J. Wang, Z. Liao, Efficiency-enhanced inverted Doherty power amplifier with class-E peaking amplifier, in *2010 International Symposium on Signals, Systems and Electronics*, vol. 2 (2010), pp. 1–3
7. M. Hashemi, L. Zhou, Y. Shen, L.C.N. de Vreede, A highly linear wideband polar class-E CMOS digital Doherty power amplifier. IEEE Trans. Microw. Theory Tech. **67**(10), 4232–4245 (2019)
8. G.W. Choi, H.J. Kim, W.J. Hwang, S.W. Shin, J.J. Choi, S.J. Ha, High efficiency class-E Tuned Doherty amplifier using GaN HEMT, in *2009 IEEE MTT-S International Microwave Symposium Digest* (2009), pp. 925–928
9. N. Elsayed, H. Saleh, M. Baker, M. Sanduleanu, A 28GHz, asymmetrical, modified Doherty power amplifier, in 22nm FDSOI CMOS, in *IEEE International Symposium on Circuits and Systems* (2020)
10. Z. Zong, X. Tang, K. Khalaf, D. Yan, G. Mangraviti, J. Nguyen, Y. Liu, P. Wambacq, A 28Ghz voltage-combined Doherty power amplifier with a compact transformer-based output combiner in 22nm FD-SOI, in *2020 IEEE Radio Frequency Integrated Circuits Symposium (RFIC)* (2020), pp. 299–302
11. E. Kaymaksut, D. Zhao, P. Reynaert, E-band transformer-based Doherty power amplifier in 40 nm CMOS, in *2014 IEEE Radio Frequency Integrated Circuits Symposium* (2014), pp. 167–170
12. B. Wicks, E. Skafidas, R. Evans, A 60-GHz fully-integrated Doherty power amplifier based on 0.13-um CMOS process, in *2008 IEEE Radio Frequency Integrated Circuits Symposium* (2008), pp. 69–72
13. S. Shopov, R.E. Amaya, J.W.M. Rogers, C. Plett, Adapting the Doherty amplifier for millimetre-wave CMOS applications, in *2011 IEEE 9th International New Circuits and Systems Conference*, (2011), pp. 229–232
14. N. Rostomyan, M. Özen, P. Asbeck, 28 GHz Doherty power amplifier in CMOS SOI with 28% back-off PAE. IEEE Microw. Wireless Compon. Lett. **28**(5), 446–448 (2018)

Chapter 6
A 28 GHz Inverse Class-D Power Amplifier

6.1 Conventional CMCD PA

Unlike linear PAs, the transistors in SMPAs act as a switch. In a voltage mode class-D (VMCD) such as in Fig. 6.1, a square input signal is applied, and an LC resonant tank is inserted in series with the load resistance to enable a sinusoidal current to pass through. This results in a non-overlapping peak drain voltage of VDD and a peak drain current of VDD/R_{load}. This non-overlapping behavior results in a 100% theoretical efficiency. However, the VMCD PA results in a large power dissipation at GHz frequencies.

The CMCD PA overcomes this by utilizing an LC tank resonating at the fundamental frequency that absorbs the parasitic drain capacitance into the output network as shown in Fig. 6.2. The output voltage is sinusoidal where the overlap between current and voltage is also minimized. The drain voltages of each transistor (V_{D1}, V_{D2}) are half-sinusoidal. The current waveforms (i_{D1}, i_{D2}) are square wave. The transistor currents i_{D1} and i_{D2} can be described as

$$i_{D1} = \frac{V_{D1}}{R_{on}} \left[\frac{1}{2} + \frac{2}{\pi} \sum_{k(odd)}^{inf} \frac{sin(k\phi)}{k} \right]$$

$$i_{D2} = \frac{V_{D2}}{R_{on}} \left[\frac{1}{2} - \frac{2}{\pi} \sum_{k(odd)}^{inf} \frac{sin(k\phi)}{k} \right]$$

(6.1)

where R_{on} is the ON-resistance of the device. The voltage on R_{load} can then be expressed as

$$R_{load} = V_{D1} - V_{D2} = Acos(\phi) + Bsin(\phi)$$

(6.2)

© The Author(s), under exclusive license to Springer Nature Switzerland AG 2022
N. Elsayed et al., *High Efficiency Power Amplifier Design for 28 GHz 5G Transmitters*, Analog Circuits and Signal Processing,
https://doi.org/10.1007/978-3-030-92746-2_6

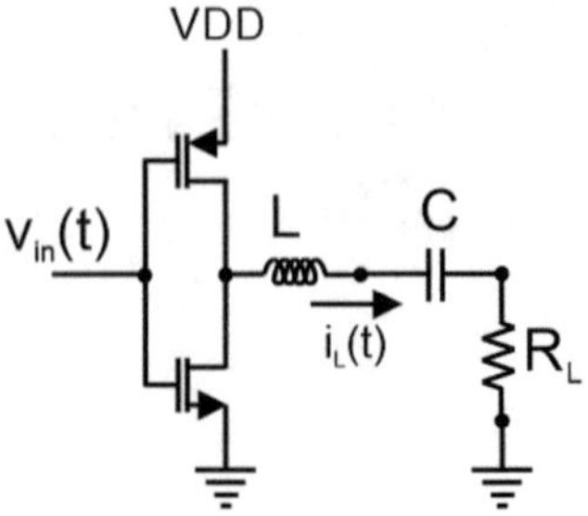

Fig. 6.1 Schematics of
VMCD PA

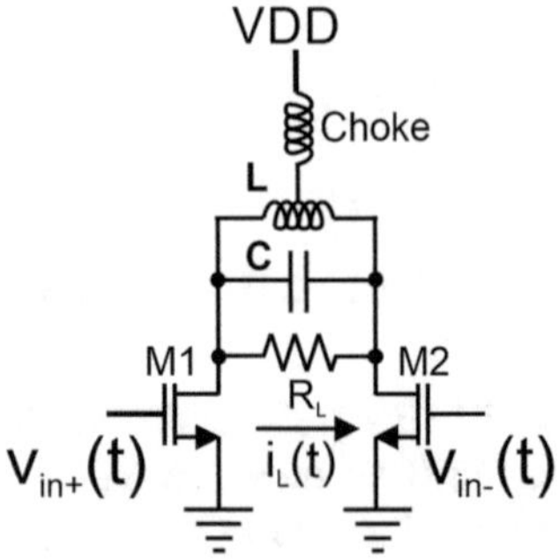

Fig. 6.2 Schematics of a
conventional CMCD PA

where A and B are the amplitudes of the two drain–source voltages with phase ϕ.
Since the RLC tank allows only the resonance frequency current through the load,
a sinusoidal voltage at f_0 is seen at R_{load} assuming $A = B$. The output power can
then be defined as

$$P_{out} = \frac{V_{out}}{2R_{load}} = \frac{A^2 + B^2}{2R_{Load}} = \frac{A^2}{R_{Load}} \tag{6.3}$$

assuming A and B are equal; and the DC current through the transistors would be
calculated as

$$I_{DC} = \frac{VDD - A/\pi}{2R_{on}} \tag{6.4}$$

The drain efficiency (DE) of the CMCD PA can then be represented by

$$\eta = \frac{P_{out}}{P_{DC}} = \frac{A^2 R_{on}}{VDD(VDD - A/\pi)} \tag{6.5}$$

From the equation above, it can be deduced that in order to increase the efficiency
R_{on} must be increased, which means a smaller device size. This then is conductive
to lower output peak power. In order to distribute the voltage stress, use a higher
supply voltage, and deliver higher output power, a cascode configuration could be
used with a relatively larger transistor size.

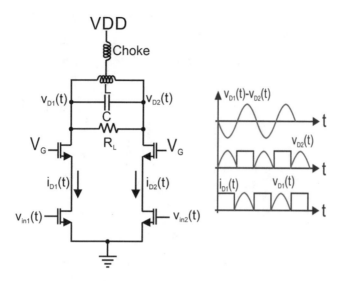

Fig. 6.3 Schematics and the ideal waveforms of the drain voltages V_{D1}, V_{D2}, I_{D1} of a cascode CMCD PA

Figure 6.3 shows the topology of a cascode CMCD PA. The two input transistors are driven by equal input signals that are 180° out of phase. The cascode transistors are biased at a constant DC input and serve the purpose of overcoming the breakdown voltage of the devices by allowing for a higher V_{DD}. The circuit is symmetrical, and the RLC tank results in a sinusoidal voltage output at the drain. Figure 6.3 shows the current and voltage waveforms of an ideal stacked CMCD PA. So an ideal CMCD PA should see an open circuit with no current for even harmonics and short circuit for all odd harmonics resembling a push–pull version of inverse Class-F PA [1] (Fig. 6.4).

Current mode (inverse) Class-D (CMCD), introduced in [2], is able to achieve the largest peak power with a theoretically 100% efficiency. Inverse Class-D PA can also provide higher bandwidth compared to other SMPAs in case a wideband input matching is used [3]. CMCD also has the advantage of having the output drain capacitance absorbed into the output matching network allowing it to operate at higher frequencies compared to other SMPAs along with a relatively simple output matching network [4]. Some works in the literature as in [4–9] have implemented CMCD PAs with various CMOS technology or III–V technologies at lower frequencies. A cascode CMCD PA in 130 nm CMOS was presented in [4]; however, its peak performance was recorded at 1.4–2.6 GHz. It utilizes integrated balun transformers that degrade the efficiency and increase the area. This work will utilize the high power capabilities of 22 nm FDSOI technology with high f_{max} of 371 GHz (NFET) at RF frequencies due to reduced parasitic to the substrate [10] along with the capability of CMCD to operate at high frequencies compared to other SMPAs.

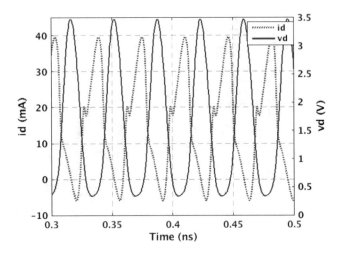

Fig. 6.4 Simulated non-overlapping drain voltage (V_{D1}) and current (i_{D1})

6.2   Proposed CMCD Topology

In order to enable the use of high output power, a higher supply voltage is needed. However, due to the limited breakdown voltage of the devices, this is limited. Thus, a cascode configuration is used.

Adding a cascode transistor, however, affects the charging and discharging time of the capacitance at the middle node negatively. This additional capacitance can be tuned out by adding a parallel inductor at the common node between the input transistors and the stacked ones. This helps achieve Zero Voltage Switching (ZVS) at high frequencies. Fixed gate bias of stacked CMCD PA suffers from efficiency degradation. The output capacitance increases by the gate–source capacitance of the cascode transistor resulting in a resonance frequency shift to a lower frequency.

In order to minimize the output parasitic capacitance, different gate biasing techniques can be used. A novel technique called pulse injection is proposed that entails the injection of signal from the input transistor to the output transistor (see Fig. 6.5). This way, we prevent the input transistor to turn on when the cascode transistor is off. Figure 6.6 shows the improvement in peak DE when using pulse injection against having a fixed gate bias at the cascode transistor by 28%.

The input RF signal is applied to the gates of the under transistor and cascode transistor through transmission lines terminated on their characteristic impedance used also for biasing the gates of the cascode and under transistor to two different gate voltages VG2 and VG1. The lines are designed to have 100 Ω so at the RF inputs we "see" 50 Ω. The inductor L2 serves to resonate out the gate–source capacitances of the transistors M2 and M4 and the parasitic drain capacitances of the transistors M1 and M3. The center-tap inductor L1 connects to the supply voltage through a transmission line in order to choke the fundamental current of the tank.

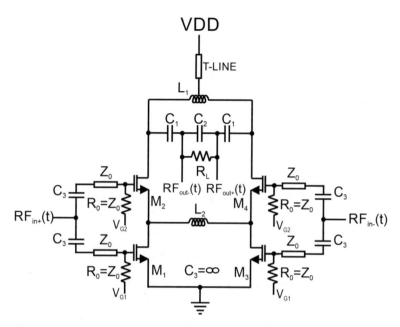

Fig. 6.5 The proposed pulse injected cascode CMCD PA

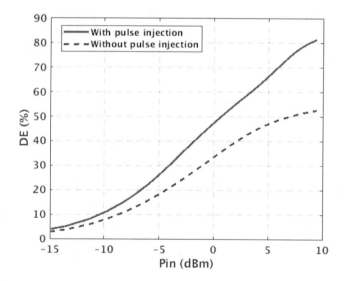

Fig. 6.6 Simulated DE for fixed gate bias of the cascode transistor and using pulse injection. Improvement in DE using pulse injection against using a fixed gate bias by 28%

The load is resonant with inductor L1 and the capacitors C1-C2-C1. The purpose of the capacitive divider is to transform the high impedance of the tank into a differential impedance of 100 Ω. This can apply to a 100 Ω differential antenna

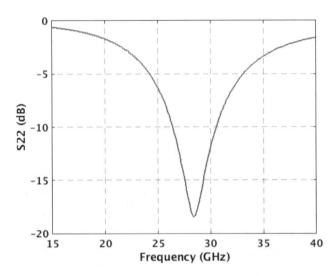

Fig. 6.7 Simulated characteristics of output matching network (S_{22}) with -18 dB at 28 GHz

or a single ended 50 Ω antenna if one of the outputs is terminated with a dummy on-chip 50 Ω resistance. Figure 6.5 shows the schematics of the pulse injected CMCD PA. Figure 6.4 shows the simulated non-overlapping drain voltage and current waveforms of the output transistor using harmonic balance simulation at peak PAE. Figure 6.7 shows the output matching of the RLC network.

6.3   Measurements Results

The chip in Fig. 6.8 was implemented in the Global Foundries' CMOS 22nm FDSOI technology. The testing was done on-chip using the Elite 300 semi-automatic probe station. The input and output signals are measured using ground-signal-ground-signal-ground (GSGSG) probes, while the DC signals are applied through a multi-wedge probe. The applied out-of-phase input signals were supplied from two signal generators frequency locked to the same reference and adjusted in antiphase (180°) from their phase control knob. The measurements took into account the losses of the connected wires. Figure 6.10 shows the simulated and measured PAE and DE.

The measured peak PAE and DE show 46%/72%, respectively (see Fig. 6.10), with peak PAE at 0 dBm input power. Figure 6.9 shows a maximum output power of 19 dBm with a power gain of 17 dB to a 50 Ω load. The frequency was swept from 26 GHz to 30 GHz and output power was recorded. The CMCD PA shows a narrow-band response as shown in Fig. 6.12 with output power greater than 15 dBm for a 1 GHz bandwidth (Figs. 6.9, 6.10, 6.11, 6.12).

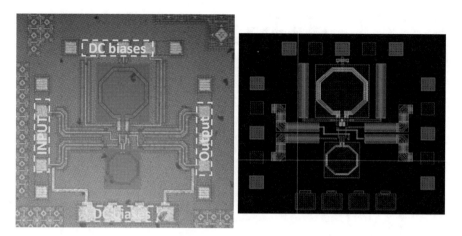

Fig. 6.8 CMCD PA chip microphotograph and layout (0.62 mm × 0.67 mm) with GSGSG probes

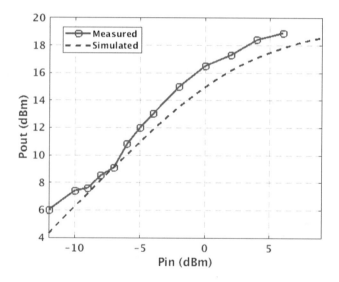

Fig. 6.9 Measured and simulated P_{out} vs. P_{in}. The CMCD PA shows a maximum output power of 19 dBm

In order to measure power delivering capabilities, the supply voltage was increased and output power was recorded (see Fig. 6.11). The CMCD PA is able to deliver higher power with a higher V_{DD}; however, it is limited to the breakdown voltage of the device.

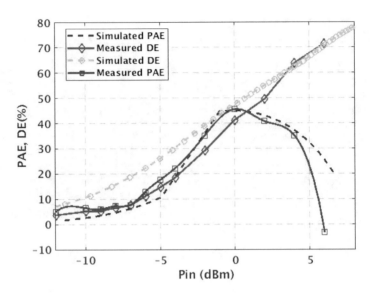

Fig. 6.10 Measured and simulated PAE and DE. The CMCD PA shows a peak PAE of 46% and a peak DE of 72%

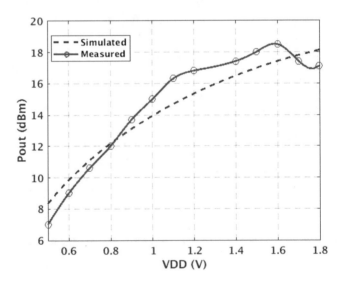

Fig. 6.11 Measured output power (P_{out}) vs. supply voltage (V_{DD}). Output power increases with supply voltage but limited to the breakdown voltage of the device

6.3.1   Comparison with State-of-the-Art CMCD PAs

Table 6.1 shows a comparison between the performance of the proposed pulse injected CMCD PA and other CMCD PAs reported in the literature in various

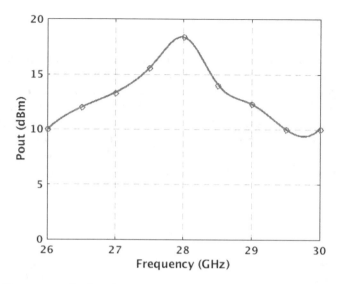

Fig. 6.12 Output power (P_{out}) vs. frequency shows a narrow-band behavior maintaining $P_{out} >$ 15 for a BW of 1 GHz

Table 6.1 Comparison between the proposed 22 nm FDSOI CMCD PA and related work in the literature

Ref.	Technology	Freq. (GHz)	P_{out} (dBm)	Peak PAE (%)	Peak DE (%)	Gain (dB)
This work	22nm FSOI CMOS	28	19	46	71	16.5
2014 [4]	130nm CMOS	1.8	26.8	45	46	N/A
2011 [5]	65nm CMOS	2.25	21.8	44.2	78	N/A
2020 [6]	180nm CMOS	1.85–1.91	17.5	17.5	N/A	18.91
2013 [7]	GaAs HBT	0.7	29.5	68.5	N/A	11
2013 [8]	GaN	2.6	29.5	62	N/A	14.4
2020 [9]	130nm CMOS[a]	2.4	19.75	58	74.1	19.5

[a] Post-layout simulation

technologies. When comparing the performance of PAs, it is important to keep in mind the operating frequency compared to the f_{max} of the devices used. We are able to report the best measured peak PAE in the literature compared to other CMCD PAs in CMOS. Our design also utilizes much less area due to a less number of passive elements, through only using 2 inductors compared to other designs using 5 such as in [6]. Other technologies such as GaAs HBT and GaN are able to report higher efficiencies and output power on the expense of cost and integration capabilities such as in [7, 8] at lower frequency bands. Overall, this design is able to achieve the best performance at RF frequencies as high as 28 GHz compared to similar work reporting compatible efficiencies at 2.25 GHz in [4, 5].

6.4   Summary

In this chapter, a 28 GHz current mode Class-D was implemented and measured in 22 nm FDSOI. In order to overcome the limited breakdown voltage of the devices, use a higher supply voltage, and deliver high output power, the architecture relies on stacked topology. Pulse injection was implemented in the cascode transistors to improve efficiency. The measured CMCD PA reports a peak PAE of 46% and a peak PAE of 71% with an output power of 19 dBm. It represents the highest performance reported for a CMCD PA in CMOS at 28 GHz. This amplifier is suitable for outphasing transmitters or a Doherty configuration to improve efficiency.

References

1. C. Schuberth, P. Singerl, H. Arthaber, M. Gadringer, G. Magerl, Design of a current mode Class-D RF amplifier using load pull techniques, in *2009 IEEE MTT-S International Microwave Symposium Digest* (2009), pp. 1521–1524
2. H. Kobayashi, J. Hinrichs, P.M. Asbeck, Current mode Class-D power amplifiers for high efficiency RF applications, in *2001 IEEE MTT-S International Microwave Symposium Digest (Cat. No.01CH37157)*, vol. 2 (2001), pp. 939–942
3. H.M. Nemati, C. Fager, H. Zirath, High efficiency LDMOS current mode Class-D power amplifier at 1 GHz, in *2006 European Microwave Conference* (2006), pp. 176–179
4. H.R. Khan, A.R. Qureshi, F. Zafar, Q. ul Wahab, Design of a broadband current mode class-D power amplifier with harmonic suppression, in *2014 IEEE 12th International New Circuits and Systems Conference (NEWCAS)* (2014), pp. 169–172
5. D. Chowdhury, S.V. Thyagarajan, L. Ye, E. Alon, A.M. Niknejad, A fully-integrated efficient CMOS inverse Class-D power amplifier for digital polar transmitters, in *2011 IEEE Radio Frequency Integrated Circuits Symposium* (2011), pp. 1–4
6. M. Harifi-Mood, A. Bijari, H. Alizadeh, N. Kandalaft, A new highly power-efficient inverse class-D PA for NB-IoT applications, in *2020 10th Annual Computing and Communication Workshop and Conference (CCWC)* (2020), pp. 0458–0462
7. T.-P. Hung, A.G. Metzger, P.J. Zampardi, M. Iwamoto, P.M. Asbeck, Design of high-efficiency current-mode Class-D amplifiers for wireless handsets. IEEE Trans. Microw. Theory Tech. **53**(1), 144–151 (2005)
8. A. Sigg, S. Heck, A. Bräckle, M. Berroth, High efficiency GaN current-mode Class-D amplifier at 2.6 GHz using pure differential transmission line filters. Electron. Lett. **49**(1), 47–49 (2013)
9. M. Silva-Pereira, M. Assunção, J.C. Vaz, Analysis and design of current mode Class-D power amplifiers with finite feeding inductors. IEEE Trans. Very Large Scale Integr. Syst. **28** 1–10 (2020)
10. S.N. Ong, S. Lehmann, W.H. Chow, C. Zhang, C. Schippel, L.H.K. Chan, Y. Andee, M. Hauschildt, K.K.S. Tan, J. Watts, C.K. Lim, A. Divay, J. S. Wong, Z. Zhao, M. Govindarajan, C. Schwan, A. Huschka, A. Bellaouar, W. L Oo, J. Mazurier, C. Grass, R. Taylor, K. W. J. Chew, S. Embabi, G. Workman, A. Pakfar, S. Morvan, K. Sundaram, M.T. Lau, B. Rice, D. Harame, A 22 nm FDSOI technology optimized for RF/mmWave applications, in *2018 IEEE Radio Frequency Integrated Circuits Symposium (RFIC)* (2018), pp. 72–75

Chapter 7
Phased-Array Transmitter

7.1 Conventional Direct Conversion Transmitter Architecture

A transmitter modulates data to a predetermined carrier frequency f_c (RF). The transmitter then has to provide power amplification through the use of PAs without distorting the original signal or causing any undesired frequency spurs in nearby channels. To find the suitable design of a transmitter, the type of modulation has to be determined first. Then optimization and trade-offs have to be kept in mind to maintain low cost and high levels of integration.

The main parts of the direct conversion RF transmitter are:

1. Local oscillator (LO): This generates a CW signal of a frequency that, mixed with the input signal, results in the desired frequency. The LO signal is synchronized to the clock reference signal [1].
2. I/Q Modulator: The modulator mixes the baseband signal with the LO signal.
3. Power Amplifier: Transforms DC power into RF power that is high enough to travel a distance over the wireless medium.
4. Band Pass Filter: Performs out-of-ban signals rejection.
5. Antenna: Provides means to transmit and receive RF signals through a wireless medium.

For this work, a direct modulation topology is used where the data is up-converted into the RF frequency directly with the use of one mixer. This architecture is less complex than its counterparts.

However, one key drawback of implementing this with scaled CMOS devices at high frequencies is the low breakdown voltage of the devices themselves, rendering the transmitter from achieving efficient high power generation. One way to address this issue is to use phased-array architectures. Phased-array architecture combines the output power from several devices through air (in case of wireless

N. Elsayed et al., *High Efficiency Power Amplifier Design for 28 GHz 5G Transmitters*, Analog Circuits and Signal Processing, https://doi.org/10.1007/978-3-030-92746-2_7

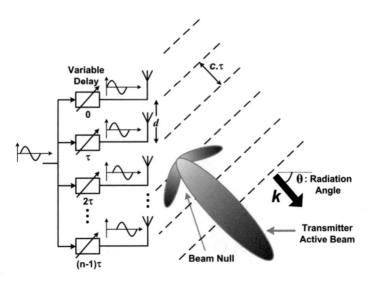

Fig. 7.1 N-phased-array transmitter [3]

communication). However, in order for this to work, a beam is formed in a desired direction by changing the relative delay in each *leg* to compensate for the propagation delays for signals from different elements. This allows for electronic control over beam steering without varying the physical orientations of the antennas. This concept enables spatial power combining usually done by utilizing phase shifters. Phased-array systems have proven to provide higher effective isotropic radiated power (EIRP) and lower system SNR [2].

Integration on silicon makes it feasible to achieve complex phased-array systems where mixed-signal and digital circuits can be on the same chip with low cost and high level of reliability. The rise of scaled CMOS technologies and the ability to integrate multiple antenna systems on silicon processes allow for more complex and cost effective solutions for high frequency communication schemes. Figure 7.1 depicts an N-phased-array transmitter with an input signal $s(t)$ distributed through N paths where the delay signal is the multiple of τ and θ is the radiation angle [3].

7.2   Proposed Phased-Array Transmitter Architecture

In Fig. 7.2 a 28 GHz phased-array transmitter for 5G is presented. The baseband signals produced by a fast D/A converter are applied to the low pass filter and the I/Q up-converter. Thereafter, the signal should be split so it can be applied to the 4RF paths with a tunable delay line, phase selector (0 or 180°) VGA, and PA. The power divider should not incur many losses and should be broadband. The polyphase filter,

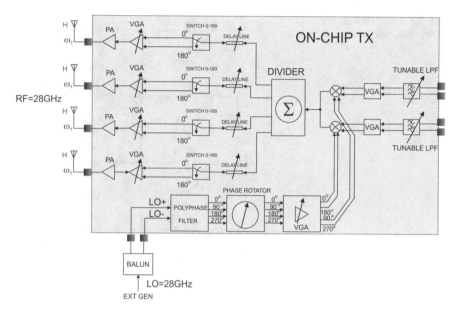

Fig. 7.2 1:4 power divider in a phased-array transmitter

phase rotator, and the tunable LPF were designed by my colleagues (Dan Cracan and Ademola Mustapha). However, these blocks will be briefly discussed to provide an understanding of the whole transmitter chain.

7.3   LO Quadrature Generation

1. **Polyphase Filter**
   The 2 mixers require 4 quadrature signals. A differential LO signal is provided externally to the polyphase filter that then produces multiple phases. The number of phases depends on the number of RC elements. A two-stage polyphase filter shown in Fig. 7.3 was designed to improve accuracy. Each stage consists of 4 RC elements in order to produce 4 phases.

2. **Phase Rotator**
   In order to have control over the LO phase and avoid phase mismatch, a phase rotator circuit shown in Fig. 7.4 is added after the polyphase filter. The phase rotator takes S_i and S_q as inputs to control the phase. I_1 and I_2 are bias currents used to provide a fine control over the phase. The LC tank was designed to resonate and provide maximum gain at the LO frequency of 28 GHz.

3. **Mixer**
   A differential topology was implemented to multiple the LO signal with the BB as shown in Fig. 7.5.

Fig. 7.3 Circuit schematics
of the polyphase filter

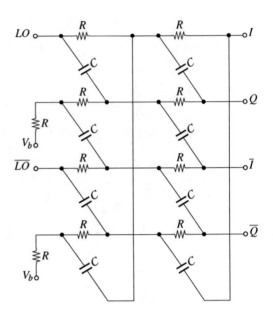

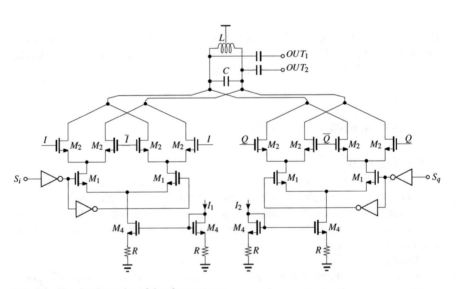

Fig. 7.4 Circuit schematics of the phase rotator

Fig. 7.5 Circuit schematics
of 28 GHz mixer.

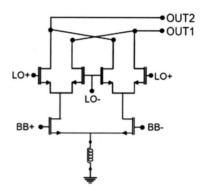

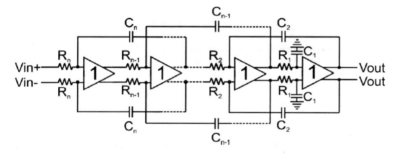

Fig. 7.6 Differential 10-stage filter

7.4 Tunable Low Pass Filter

A seven-stage differential all-pole filter based on unity gain buffers is used in the
transmitter design. The transfer function can then be defined as

$$H(s) = \frac{1}{1 + sRC + s^2 R^2 C^2 + \ldots + s^n R^n C^n} \tag{7.1}$$

In this equation, it is assumed that the buffer has infinite input impedance and zero
output impedance. The transfer function only contains poles where more stages
could be added to change the order of the filter. Since a sharp roll-off is required, a
10-stage filter was implemented that has differential topology for improved common
mode rejection shown in Fig. 7.6.

The cut-off frequency of the filter is determined by the resistor and capacitor
values. Hence, to enable bandwidth tuning, MOS transistors in triode were utilized
as voltage controlled capacitor. Two controls were used: coarse and fine tuning,
by switching on a parallel resistor and changing the MOS resistance, respectively.
Figure 7.7 shows the S-parameters of the filter with bandwidth tuning between 0.7
and 1.5 GHz.

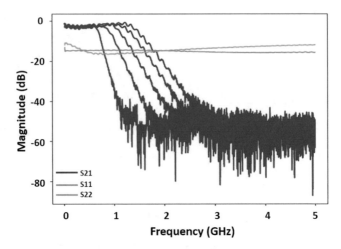

Fig. 7.7 Measured S-parameters of 10-stage tunable filter. The figure shows varying cut-off frequencies with varying settings exhibited in S21 with a constant S11 and S22

7.5 Low Frequency VGA

A VGA shown in Fig. 7.8 on the baseband level was designed in order to have a form of control on the signal level. The VGA takes the differential signal V_{in} and produces the out-of-phase signals $Out-$ and $Out+$. The gain of the VGA is determined by the inputs $I1$ and $I2$.

The values of $I1$ and $I2$ are controlled by a 6-bit input digital to analog converter shown in Fig. 7.9. The DAC is controlled by an input current connected to the source of 6 nFETs of which the size is scaled according to the binary weight. Each of these is then stacked with two transistors of identical size whose gate is controlled by a bit and its complement. The sum of the currents produced by each transistor (Iout1, Iout2) are then used as control input currents to the LF VGA (I1, I2).

Figure 7.10 shows the frequency response of the low frequency VGA against frequency with different control bits from the DAC. A maximum gain of 4 dB is achieved when all the 6 bits are at the highest level (6'b11111). Intermediate gain values can be achieved by adjusting the control bit values. A bandwidth of 1.6 GHz is also realized, which is sufficient for a baseband signal between 100 MHz and 1 GHz.

7.6 Power Divider Design

The starting point for the proposed power divider is the active power divider with phase inversion from Fig. 7.11. The transistor M0 has an equal load Z_0 in the source and drain. As a result the input RF signal is amplified in anti-phase at the drain with

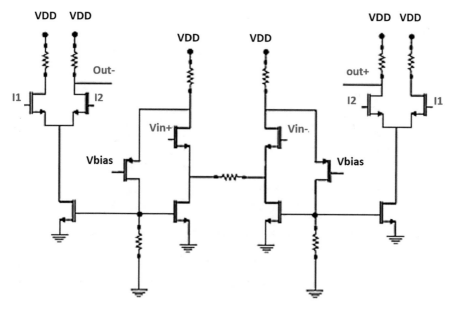

Fig. 7.8 Low frequency VGA

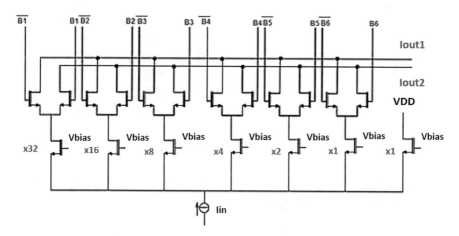

Fig. 7.9 6-bit analog-to-digital converter

a gain of -1 and amplified in phase with gain $+1$ at the source. The drain and the source are loaded with transmission lines of impedance Z_0. As long as the phase inversion is not important, this power divider is very simple (Fig. 7.12). However, if we need slight power gain, we need a different approach. In Fig. 7.13, a simple gain stage with a cascode amplifier is presented. The input RF signal is transformed into a current i_D by the transconductance of the transistor M0. The current i_D is then

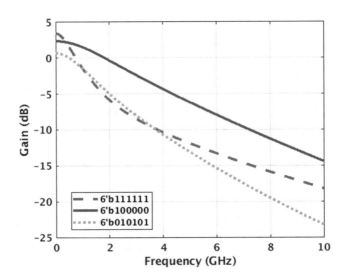

Fig. 7.10 VGA gain with different control bit levels

Fig. 7.11 1:4 Wilkinson power divider from 3 1:2 power dividers

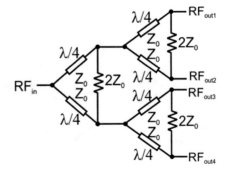

transformed into a voltage again at the drain of transistor M1 with a voltage gain of $\breve{g}_{m0}R_0$ thus allowing for power gain as shown in Eq. 7.2.

$$G = 10\log\frac{P_{out}}{P_{in}} = \frac{\frac{V_{out}^2}{R_0}}{\frac{V_{in}^2}{R_0}} = 20\log\frac{V_{out}}{V_{in}} = 20\log(g_m R_0) \qquad (7.2)$$

Thereafter the cascode stage can be segmented into 4 equal transistors M1–M4 with their W a quarter of the size of W of transistor M0 as shown in Fig. 7.14. As a result of the segmentation of the cascode transistor, the current i_D is divided into 4 equal replicas and transformed into a voltage at the drains of transistors M1–M4. The power gain at each four outputs is

Fig. 7.12   1:2 active power divider with phase inversion

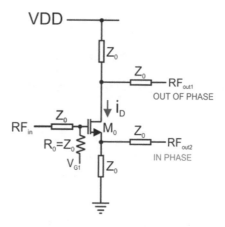

Fig. 7.13   Cascode amplifier as a basic building block for power divider

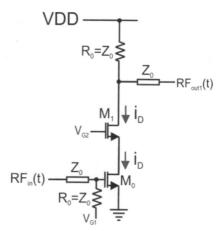

$$G = 10 \log \frac{P_{out}}{P_{in}} = \frac{\frac{V_{out}^2}{R_0}}{\frac{V_{in}^2}{R_0}} = 20 \log \frac{V_{out}}{V_{in}} = 20 \log \left(\frac{g_m R_0}{4} \right) \qquad (7.3)$$

Compared to the cascode stage from Fig. 7.13, the power gain is diminished by $20 \log 4 = 12$ dB. As long as $g_m R_0/4$ is large enough, the active power divider can produce power gain.

7.6.1   Noise Analysis for Power Divider

For the noise analysis of the 1:4 power divider, we use the conceptual Fig. 7.15. The circuit is equivalent to a noiseless power divider with an equivalent noise voltage

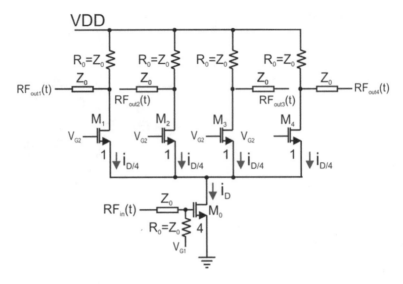

Fig. 7.14 1:4 active power divider with cascoded outputs

Fig. 7.15 Conceptual
diagram for the equivalent 1:4
noiseless power divider

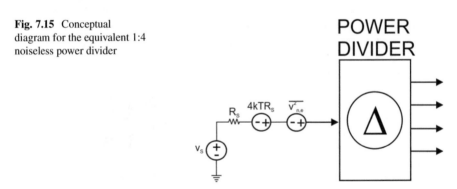

in series with the input. If R_S is the series resistance of the source, the noise figure
NF is

$$\text{NF} = 1 + \frac{\overline{v_{n,e}^2}}{4kT R_s} \tag{7.4}$$

The next step is to calculate the equivalent noise source of a cascode stage from
Fig. 7.16. The noise sources for M0 and M1 are the gate noise voltages obtained
from transforming the drain noise current into a voltage noise source at the gate.
The load resistor R_L is also generating noise, with an equivalent of $4kT R_L$. All the
noise sources are transformed into the input as

$$\overline{v_{n,eq}^2} = \overline{v_{n,e0}^2} + \frac{\overline{v_{n,e1}^2}}{(g_{m0}r_{0,0})^2} + \frac{4KT R_L}{(g_{m0}R_L)^2} \simeq \overline{v_{n,e0}^2} + \frac{4KT}{g_{m0}^2 R_L} \tag{7.5}$$

Fig. 7.16 Cascode stage with
noise sources

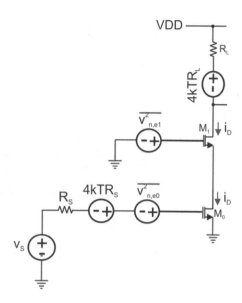

where $r_{0,0}$ is the output resistance of the transistor M0. From this we could calculate
the equivalent noise source of the 1:4 active power divider from Fig. 7.17 as

$$\overline{v_{n,e}^2} \simeq \overline{v_{n,0}^2} + \frac{4^4 KT}{g_{m,0}^2 R_L} \tag{7.6}$$

Based on Fig. 7.15 one could calculate the noise figure (NF) of the 1:4 power divider
as

$$NF_4 \simeq 1 + \frac{8KT}{3g_{m,0} R_s} + \frac{4^4 KT}{g_{m,0}^2 R_S R_L} \tag{7.7}$$

Based on the current division factor n for a 1:n generalized active power divider,
the noise figure (NF) of a 1:n active power divider from Fig. 7.14 is

$$NF_n \simeq 1 + \frac{8KT}{3g_{m,0} R_s} + \frac{4n^2 KT}{g_{m,0}^2 R_S R_L} \tag{7.8}$$

Hence, the noise figure of the 1:n power divider increases quadratically with the
number of stages n and the power gain:

$$G = 10\log \frac{P_{out}}{P_{in}} = \frac{\frac{v_{out}^2}{R_0}}{\frac{v_{in}^2}{R_0}} = 20\log \frac{V_{out}}{V_{in}} = 20\log(\frac{g_m R_0}{n}) \tag{7.9}$$

Obviously the power gain of the 1:n active power divider is reduced when increasing
the number of stages and its noise figure is increased.

VDD

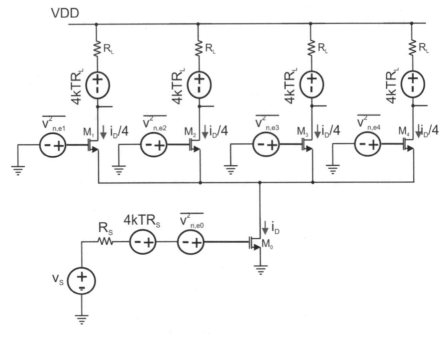

Fig. 7.17 Active 1:4 power divider with noise sources

7.6.2   Simulations and Measurement Results

The 1:4 power divider was realized in the 22nm FDSOI CMOS Technology from Global Foundries and its chip photomicrograph is shown in Fig. 7.18. In Fig. 7.19 the measured and simulated input return loss S_{11} is shown. The measurements have been carried out across 8 different chips. It shows broadband matching around 28 GHz, and for some chips the measurements are better than simulations. Figure 7.20 presents the measured/simulated output return loss S_{22} versus frequency measured across 8 different chips. It shows quite good correlation between measurements and simulations. In Fig. 7.21 the measured/simulated DUT power gain vs. frequency is shown.

As expected, the proposed active power divider exhibits a slight power gain of 1–2 dB and the simulations match quite well the measured results. The measured/simulated power gain vs. input power Pin is presented in Fig. 7.22. As expected the power gain will compress at higher input power levels, and the simulations predict well the behavior of the power gain. For estimating the 1 dB compression point, Fig. 7.23 shows the measured/simulated output power, Pout versus input power Pin for four different chips.

The 1-dB compression point is around 7 dB input power. The broadband noise figure of the power divider was measured using the Anritsu 125 GHz Vector Star

Fig. 7.18 Chip
photomicrograph
(700×600um)

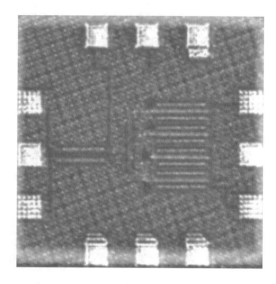

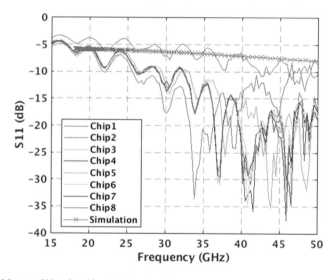

Fig. 7.19 Measured/simulated input return loss S_{11}

Analyzer using the concept of down-converting the signal to baseband and digitizing
the signal using an A/D converter without the need of a calibrated noise source.
The result measured across 6 chips, together with the simulated NF, is shown in
Fig. 7.24. As shown in Fig. 7.24, the measured results have a higher noise figure
than the simulated values. For estimating the non-linearity of the power divider we
applied two tones at 19.9 and 20.25 GHz, respectively. The power of the two input
tones at the input of the DUT is Pin= -2.8 dBm. The two tones, together with

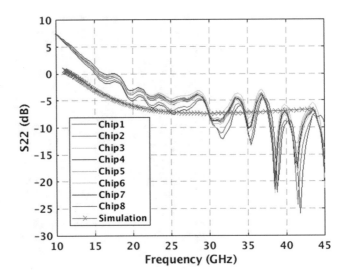

Fig. 7.20 Measured/simulated output return loss S_{22}

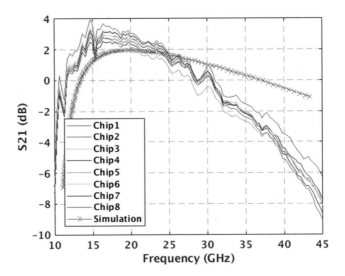

Fig. 7.21 Measured/simulated DUT power gain S_{21}

the third order intermodulation distortion, are shown in Fig. 7.25. The difference
between the two main tones is $\Delta = 20\,\text{dB}$. According to the well known formula:

$$IIP3 = \text{Pin} + \frac{\Delta}{2} = -2.8 + \frac{20}{2} = +7.2\,\text{dBm} \qquad (7.10)$$

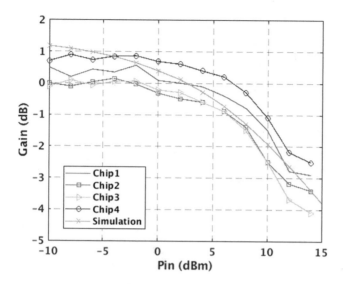

Fig. 7.22   Measured/simulated power gain vs. input power Pin

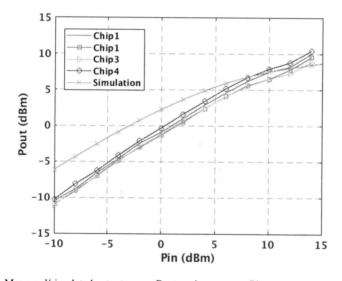

Fig. 7.23   Measured/simulated output power Pout vs. input power Pin

7.6.3   Comparison with State of the Art

Table 7.1 shows comparison with previously reported active power splitter in CMOS. In [4], a 57–65 GHz inductor-less active 1:4 power splitter in 90nm LP CMOS was implemented for phased-array transmitter implemented in as cascode

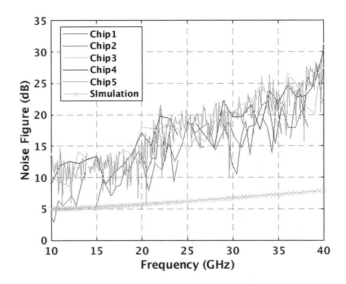

Fig. 7.24 Measured/simulated noise figure NF vs. frequency

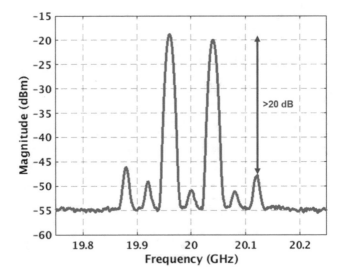

Fig. 7.25 Measured intermodulation distortion IIP3

topology. This design reports > 0 dB gain and relatively smaller area than other designs and, however, still suffers from low linearity.

In [5], a bidirectional power divider/combiner in 130nm SiGe BiCMOS was proposed with a bandwidth of 2–22 GHz. The divider records high gain (9.6 dB); however, power consumption is a major issue (100 mW), along with a very large chip area due to the presence of 10 inductors in the design. An interleaved

Table 7.1 Performance summary and comparison with the state-of-the-art CMOS power dividers

Ref.	This work	[4]	[5]	[6]	[7]	[8]
Year	2020	2012	2019	2015	2014	2010
Technology	22nm FDSOI	90nm LP CMOS	0.13um CMOS	90nm CMOS	65nm CMOS	0.18um CMOS
Type	Active	Active	Active	Active	Active	Active
1:N	1:4	1:4	1:2	1:2	1:2	1:2
Frequency (GHz)	15–35	57–65	2–22	0–40	36–46	0–20
Gain (dB)	1.1	<0	9.6	6	7	5
OP1dB (dBm)	6	−2.67	3.4	5	5	0
IIP3 (dBm)	7.2	N/A	N/A	N/A	N/A	N/A
NF (dB)	15	N/A	N/A	6	N/A	9.4
Chip area (mm×mm)	0.7×0.6	0.62×0.86	1.43×0.92	1.16×0.74	N/A	1.1×1.1
DC power (mW)	29	40	100	210	96	160

transmission line gain cells divider was proposed in [6] in 90nm CMOS in contrast with the widely used parallel topology. The divider makes use of 5 gain stages and achieves a measured power gain of 5 dB. DC power consumption, however, is very large (210 mW), and the design includes 20 inductors making the divider occupy a large area, which is impractical for a phased-array transmitter design.

In [7], a transformer-based power splitter along with a three-stage common source amplifier was implemented. This work presents a 1:2 highly linear 6 dB gain power splitter at 38 GHz in 65nm; however, the chip area is not reported. Given the presence of transformers, the power consumption is substantial if the divider were to be expanded into a phased-array Tx application. The proposed 1:4 active power divider exhibits the smallest chip area due to the absence of inductors compared to other works. The proposed divider occupies less area, less power and exhibits a more linear behavior making it ideal for compact transceiver design. It also consumes the least power (29 mW) compared to the state-of-the-art designs. This design can be used in a phased-array transceiver without consuming too much power, area, or causing any losses.

7.7 Simulation Results

Figure 7.26 shows the layout of the 4-phased-array transmitter, and Fig. 7.27 shows the microphotograph of the 22nm FDSOI chip. Periodic Steady State (PSS) post-layout simulation results with RC extraction were performed. Because of the multiple input/output pads, on-chip measurement is not possible. Currently, a PCB design is in progress to perform validation through measurement for the transmitter chain.

The transmitter linearity is measured through 1-dB compression point (see Fig. 7.28). Input inferred 1-dB compression point is −1 dBm. As another measure

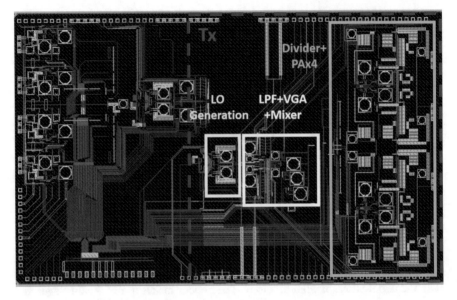

Fig. 7.26 4-phased-array transceiver layout

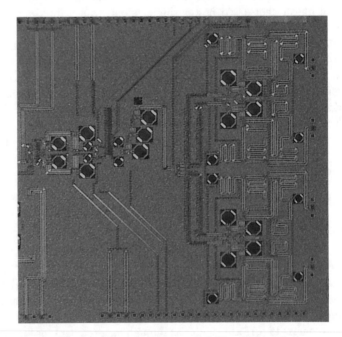

Fig. 7.27 4-phased-array transmitter microphotograph (3.75×3.824 mm).

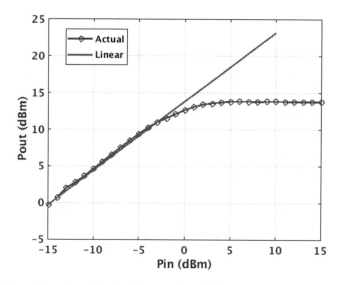

Fig. 7.28 P_{in} vs. P_{out} of one branch of the transmitter chain

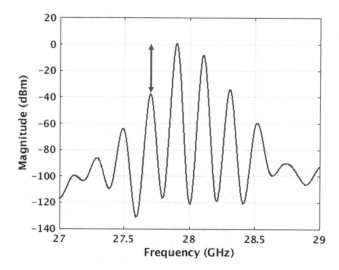

Fig. 7.29 Simulated third order intermodulation distortion with two tones (0.9 and 1.1 GHz).

of linearity, IIP3 was simulated. Two tones, 0.9 and 1.1 GHz, were applied at the input at an input power of -15 dBm for both. The two tones and third order intermodulation distortion are shown in Fig. 7.29. The difference between the two tones is $\delta = 40$ dB. To calculate the IIP3:

$$IIP3 = P_{in} + \frac{\delta}{2} = -15 + \frac{40}{2} = 5 \text{ dBm} \tag{7.11}$$

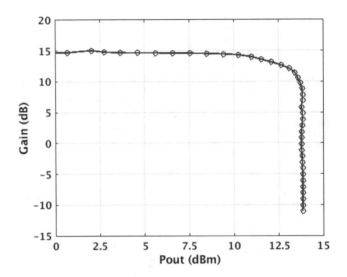

Fig. 7.30 Conversion gain of one branch of the transmitter chain

The ratio of the desired output signal power to the input signal power (conversion power gain) was simulated as shown in Fig. 7.30 for one branch of the transmitter chain. A gain of 15 dB was observed with a saturated output power of 14 dBm.

7.8   Summary

In this chapter, a 28 GHz 4-phased-array transmitter based on Doherty Class-E power amplifier was implemented in 22nm FDSOI. The transmitter takes the signals from a fast DAC converter that are then passed through a tunable low pass filter to remove any harmonics from the output signal of the DAC. A VGA is then used to adjust the signal swing amplitude to ensure the mixer is operating at the optimum conditions. After that, the baseband signal is up-converted to the RF frequency. The signal is then applied to 4 RF paths through an active power divider. Each signal path has a tunable transmission line for and a phase selector that is then applied to the Class-E based Doherty PA. The transmitter achieves 15 dB conversion gain, with a -1 dBm compression point in post-layout simulation. Simulation results will be verified with measurements after PCB design is complete.

References

1. M. Mansour, R.S. Ghoname, A. Zekry, Building radio frequency transmitter for LTE user equipment. Commun. Appl. Electron. **2**(4), 1–9 (2015)

2. A. Natarajan, A. Komijani, X. Guan, A. Babakhani, A. Hajimiri, A 77-GHz phased-array transceiver with on-chip antennas in silicon: transmitter and local LO-path phase shifting. IEEE J. Solid-State Circuits **41**(12), 2807–2819 (2006)
3. A. Natarajan, A. Komijani, A. Hajimiri, A fully integrated 24-ghz phased-array transmitter in CMOS. IEEE J. Solid-State Circuits **40**(12), 2502–2514 (2005)
4. I.-C. Chang, J.-C. Kao, J.-J. Kuo, K.-Y. Lin, An active CMOS one-to-four power splitter for 60-GHz phased-array transmitter, in *2012 IEEE/MTT-S International Microwave Symposium Digest* (2012), pp. 1–3
5. I. Song, M.-K. Cho, J.-G. Kim, G. Hopkins, M. Mitchell, J.D. Cressler, A 2–22 GHz wideband active bi-directional power divider/combiner in 130 nm SiGe BiCMOS technology, in *2016 IEEE MTT-S International Microwave Symposium (IMS)* (2016), pp. 1–4
6. C. Huang, R. Hu, DC-40GHz wideband active power splitter design with interleaved transmission-line gain cells, in *2015 Asia-Pacific Microwave Conference (APMC)*, vol. 1 (2015), pp. 1–3
7. H.T. Duong, H.V. Le, A.T. Huynh, E. Skafidas, An active 38 GHz differential power divider for automotive radar systems in 65-nm CMOS, in *2014 1st Australian Microwave Symposium (AMS)* (2014), pp. 31–32
8. J. Huang, H. Wu, R. Hu, C.F. Jou, D. Niu, A DC-20GHz CMOS active power divider design, in *2010 Asia-Pacific Microwave Conference* (2010), pp. 524–526

Index

Printed in the United States
by Baker & Taylor Publisher Services